'Population Control' through Nuclear Pollution

'Population Control'

Foreword by Paul R. Ehrlich

Most of the blame is placed on the Atomic Energy Commission structure, which is both a promotional and regulatory agency. The AEC, the authors maintain, is interested primarily in benefits (electric power), and does not sufficiently consider costs (corpses). Unfortunately, it may be many years before the corpses can be counted because of the delayed or time-bomb action of radiation on the human body.

The authors point out that the AEC is following a pattern established by the "New Technology." Technologists invent products and services, create a demand, and then, with the help of a Madison Avenue type of hucksterism, equate it with need. Bigness is virtue and "goodness is up." Familiar examples are the nearly 100 million cars and trucks that pollute the air and gobble up vast chunks of land for expressways; greedy contractors who despoil scenic wonders to build look-alike homes; detergent manufacturers who destroy lakes and streams; and tankers which kill fish and wildlife with oil slicks. Now we have what is probably the deadliest of all pollutants in our midst—nuclear power plants — which may be infesting their own customers with potential cancers and leukemias!

What is needed is an awakening of moral and social responsibility among scientists. Placed on pedestals as heroes by an admiring public, scientists must end their self-worship and think more of the critical problems facing our country—poverty, alienation and health. Too prone to "go along" with their employers—usually universities and laboratories financed by the government or a foundation—many scientists remain silent for fear of reprisals, or because they are interested only in their specialty, or because they are complacent. Few are willing to become involved, to risk "trouble."

In our nuclear world, scientists must — more than ever before—practice self-examination. They must become adversaries who ask probing questions about crucial issues and think in terms of human values. They must broaden their horizons. The first problem, the authors point out, is "keeping the by-products from intersection with the biosphere—man, animals, plants." Scientists are also admonished "to end the cruel illusion that if environmental deterioration threatens, science will come to the rescue."

Drs. Tamplin and Gofman now offer a plan for solving this new dilemma, but it requires wide public support. The plan will not deter scientific progress, and it may well save the planet. Therefore, this relevant book should be read by everyone interested in keeping the human race alive.

through Nuclear Pollution

by Arthur R. Tamplin
and John W. Gofman

NELSON-HALL COMPANY • CHICAGO

This book is dedicated to

Linus C. Pauling,
Ralph Nader, and the
late Rachel L. Carson

Standard Book Number 911012-10-9
Library of Congress Catalog No. 77-141492

Copyright © 1970 by Arthur R. Tamplin
and John W. Gofman

All rights reserved. No part of this book may
be reproduced in any form without permission
in writing from the publisher, except by a
reviewer who wishes to quote brief passages
in connection with a review written for inclusion
in a magazine or newspaper.

Nelson-Hall Co., Publishers, 325 W. Jackson
Blvd., Chicago, Ill. 60606

Manufactured in the United States of America

Contents

Foreword *vii*
Preface *xiii*
1 Why? *1*
2 Biological effects of radiation *7*
3 Beware the Gallant Knight
   of Technological Progress *28*
4 Protection policy against future pollution *42*
5 Lip service to the public health *52*
6 Tragedy on the Colorado Plateau *128*
7 Nuclear reactors *142*
8 Undisposable radioactive waste *170*
9 Plutonium: Public health
   and technological arrogance *177*
10 The nuclear weapons program *190*
11 Moral and social responsibility
   of science and scientist *200*
12 The urgent need for scientific adversaries *228*
   Appendix: Poverty in the United States *236*
   Notes *240*

Foreword

In a scenario* published before I had seen the manuscript now before me, I speculated about the significance of the work of Drs. Gofman and Tamplin as history may determine it. I was familiar enough with their studies on radiation standards at that time—through their technical publications and documents supplementing their testimony before the Joint Committee on Atomic Energy—to be both serious (and hopeful) in referring in my scenario to "the Gofman-Tamplin Act of 1976" and the untold suffering to be averted by enforcing such restrictions as they propose. This book has reinforced both my concern over the issue of radioactivity in the environment and my admiration for Drs. Gofman and Tamplin in their courageous campaign to carry the case to the public.

The book is much more than a layman's introduction to hazards of low-dose radiation, although it fills that role. It is also a penetrating study of what is wrong with present-day "technology assessment"—the procedures (or lack of them) by which governmental agencies and selected members of the scientific, engineering, and business communities weigh the costs and benefits of new technology. The radiation standards controversy serves as a timely and revealing case study of the politics, pseudo-science, and Madison Avenue hucksterism practiced by the promoters of threatening technologies of every stripe—from persistent pesticides to strategic weapons systems. The book provides special insight, of course, into the response of the

*"Looking Backward from 2000 A.D.," *The Progressive,* April 1970.

promotional-regulatory bureaucracy to criticism from within, here in the form of Gofman's and Tamplin's work at the AEC-supported Lawrence Radiation Laboratory. Some of the antics described here would be comical, if the underlying issues were only less serious. The book is a must for students of science and public policy on these grounds alone.

The threat to health from the low doses of radiation with which the authors concern themselves may be among the most insidious faced by man. As thoroughly as Drs. Gofman and Tamplin have covered the subject, there are several aspects of this threat which cannot be stressed too much. The first is that the biological consequences of radiation exposure may first appear only many years after the actual dose is received. This phenomenon gives a great advantage in any "cost-benefit" analysis to the atomic energy promoter; the benefits (e.g., electric power) will be received immediately, but the costs (corpses) may not appear for decades. By the time the bodies do appear— in sufficient numbers that the cause is evident despite inadequate statistics and a wealth of other possible culprits—the damage is irreversible; millions, perhaps hundreds of millions, already will have received doses which cannot be undone.

A second aspect worth belaboring is the concentration of many radioactive substances in food chains. Atomic Energy Commission publicists often give the impression that one gallon of poison added to a billion gallons of river makes only 1 part per billion of poison. Unfortunately, the reconcentration of some of the most dangerous materials by filter-feeding organisms and their further concentration as these, in turn, are consumed by larger creatures, exposes man and other animals at the top of food chains to far less innocuous doses. The same argument undermines the AEC's contention that, by applying their standards at the perimeter of power plants, they are being very conservative. In fact, the doses measured there may not be the highest to which the public is exposed.

Finally, Drs. Gofman and Tamplin do a particularly great service in emphasizing the *wide variety* of peaceful uses of the atom which will contribute to the public's exposure. It is not the dose to the public from any single nuclear power plant which is likely to reach the present guideline levels, but rather the dose from *all* sources combined. Fuel reprocessing plants, uranium mine tailings, medical X-rays, atomic explosions for mining and engineering feats, even the re-entering isotope power package used in space (the safety of which Tamplin so futilely challenged) will all contribute, and it is the total which counts.

In their research and in this book, Drs. Gofman and Tamplin have concerned themselves mostly with the potential costs of radiation to the individuals exposed, in the form of induction of various kinds of cancers. As they point out, however, the hazard is compounded because radiation does genetic damage as well; this shows up as stillbirths and defective children in generations to come. In this case, too, it will be too late to undo the damage by the time the costs become fully apparent. Geneticist and Nobel Laureate Joshua Lederberg recently estimated that medical costs arising from such genetic defects might amount to $10 billion per year in a population of 300 million people exposed to 60% of the present guideline level of radiation (although, he points out, this estimate could be considerably in error in either direction).

In the face of the particularly insidious features of the radiation threat enumerated here, one might expect special caution from the agency empowered to set and enforce standards for public exposure. As this book so clearly demonstrates, that has not been the case. The initial response of the Atomic Energy Commission to the suggestion that their standards were in need of review was indignation and stock denials. While Drs. Gofman and Tamplin have persisted in assembling more and more data in support of their position, the AEC has addressed itself principally to the style and tone rather than the substance of these arguments. They have attacked the motives of Gofman and Tamplin, the word "irresponsibility" has been bandied about (as it so often is by those opposed to constructive action which threatens the status quo), and they have introduced endless "red herring" arguments about the hazards of other sources of energy and other human activities.

None of these issues are relevant to the questions at hand, which boil down to these: What is a working number for the cost in lives of a given level of radiation exposure? Shouldn't the public have some say as to what benefits, what expediency, what economic "demands" justify this particular loss of life? Shouldn't the promotional and regulatory functions for *any* technology be vested in entirely separate agencies?

I am not an expert in the biological effects of radiation, but I must assume that if the AEC had a case against the *numbers* Gofman and Tamplin have produced, they would be presenting that case. They have not done so. (They regularly *imply* that the case exists, of course.) In fact, the AEC argument seems to proceed in three stages, depending on how hard they are pressed. Stage I: "Low levels of radiation are probably harmless." Stage II: "Well, the benefits exceed the costs." Stage III: "We don't know exactly what the costs are, but to use a conservative estimate will impede Progress."

The only truth evident in any of these statements is that the exact costs are not known. There is simply not enough data on human radiation exposures to make completely accurate predictions, and animal data have many shortcomings for assessing human effects. Drs. Gofman and Tamplin have done an admirable job in wading through the data which *are* available to come up with the working number which is so badly needed for cost-benefit analysis. As befits matters of public health, they have used the most conservative assumptions consistent with the available information. And they rightly insist that the burden of *proof* that radiation is *not* as dangerous as the conservative model indicates must rest with the promoters. This is hardly a revolutionary idea, but the AEC is resisting it vigorously. They, and so many other promoters of potential environmental disasters, would apparently prefer to carry out what amounts to the ultimate experiment—global distribution of any number of insidious pollutants, with virtually all of us as unwitting guinea

Foreword xi

pigs. The behavior of the AEC in the Gofman-Tamplin case is the last in a long line of incidents clearly indicating that the Commission should be dismantled and its promotional and regulatory functions distributed to more responsible agencies.

In conclusion I would like to make a disclaimer: I am known as an advocate of population control, but I hardly regard nuclear pollution (or any other of the technological/environmental threats to which the analysis in this book so often applies) as the method of choice!

Stanford University PAUL R. EHRLICH*

---

*Dr. Paul R. Ehrlich is Professor of Biology and Director of Graduate Study for the Department of Biological Sciences at Stanford University. A specialist in population biology, he is author of the widely-known book, *The Population Bomb;* and with his wife, Anne H. Ehrlich, the new book, *Population, Resources, Environment: Issues in Human Ecology.*

Preface

Men must solve the environmental crisis which threatens our planet. We are optimistic that this can be accomplished, provided men know the dimensions of the problem they face. The first step is to understand how we have arrived at the present sorry state of affairs. Then we must appreciate the dynamic which operates against solution of the environmental crisis. Finally, we can come to grips with procedures that can lead us out of the morass that has been created.

Much of this book deals with atomic technology and its associated irreversible nuclear pollution of the planet. In large measure this emphasis is the result of our personal experiences. We learned about the nature of the environmental crisis through our own experiences in endeavoring to protect the public health with respect to nuclear pollution. The lessons to be learned from nuclear pollution are *general* lessons, applicable with equal force to every aspect of the environmental crisis generated by modern science and technology.

We are indeed hopeful that our experiences and observations may lead to constructive solutions of the critical environmental problems facing the human species.

ARTHUR R. TAMPLIN
JOHN W. GOFMAN

1 Why?

... Whether men will be able to survive the changes of environment that their own skill has brought about is an open question. If the answer is in the affirmative, it will be known some day; if not, not. If the answer is to be in the affirmative, men will have to apply scientific ways of thinking to themselves and their institutions. They cannot continue to hope, as all politicians hitherto have, that in a world where everything has changed, the political and social habits of the eighteenth century can remain inviolate. Not only will men of science have to grapple with the sciences that deal with man, but—and this is a far more difficult matter—they will have to persuade the world to listen to what they have discovered. If they cannot succeed in this difficult enterprise, man will destroy himself by his halfway cleverness. I am told that, if he were out of the way, the future would lie with rats. I hope they will find it a pleasant world, but I am glad I shall not be there.*

—BERTRAND RUSSELL

This particularly appropriate quotation from one of the world's most eminent scientist-philosophers expresses clearly and succinctly why we are writing this book. We do not wish to prepare the way for rats. But as Lord Russell has stated so well, it is crucial to persuade the world to listen to what men of science have discovered concerning the sciences that deal with man.

Such sciences provide essential messages for every living human, for at stake are two of man's greatest treasures, the health of his own generation and the health and very existence of future

*From a section in *Scientists As Writers,* ed. James Harrison (Cambridge, Mass., The M.I.T. Press, 1965), pp. 145-146.

generations. Atomic, or nuclear, energy programs, as currently conceived and burgeoning rapidly, both in the so-called peaceful and military forms, provide clear and present dangers to both of these health treasures. A characteristic of modern technologies, devoted to self-perpetuation and aggrandizement, is that they are quick to resort to the cliché that only the "expert" can understand the hazards of the technology. The public, it is stated, must place its trust and faith in these "experts."

We have far, far greater faith in the intelligence of the layman in the "public." And since it is the health of the layman and his descendants that concerns us, we are determined to explain in readily intelligible terms how nuclear technology poses grave threats to health.

Public concern is indicated

Almost no family has escaped the suffering and the tragedy of a premature death due to cancer or leukemia. Atomic radiation threatens a major increase in the occurrence of such leukemia and cancer tragedies. That people appreciate the depth of suffering occasioned by these fearsome diseases is attested to by the public generosity in supporting efforts to mitigate such suffering through medical research. Extensive, dedicated research is vigorously pursued with the goal of adding 6 months, a year, or a few years of enjoyable life for the involuntary victims of cancer or leukemia. Surely, if the public generously supports *these* efforts, it also is entitled to know that nuclear technology, unless curbed, promises to add tens of thousands of additional cancer and leukemia deaths *every* year in the United States alone. No "expert" knowledge has ever been required by the family of a cancer or leukemia victim to appreciate the meaning of these diseases.

Cancer and leukemia are major health hazards posed by atomic radiation for the generation of humans currently alive.

But there are far greater health hazards of atomic radiation, the so-called genetic hazards. Radiation delivered to the ovary of women and the testis of men damages the germ plasm which will later provide future generations of human beings, and it does so in an irreversible manner. The birth of a child handicapped by a severe physical or mental deformity has marred the life of countless families. This would be sufficient reason for extreme concern on the part of the public. But the problem is far more serious than this.

As a result of medical discoveries of recent years, it is known that almost all major diseases afflicting man are in part caused by damage to the genetic material of the ova or sperm. Thus, we now appreciate that the major killing disease of our era, namely coronary heart disease (with its attendant heart attacks) must be regarded as a genetic disease. Additionally, diabetes mellitus, rheumatoid arthritis, and schizophrenia have now been definitively added to the list of diseases known to have a genetic factor in their causation. Many medical authorities believe the further developing evidence will confirm a genetic component for almost all human disease. Schizophrenia is the major mental illness, and we must be reminded that mental illness accounts for approximately one out of every two hospital beds in our country. Nuclear technology, unless curbed, promises to add more than a hundred thousand extra cases of all types of genetically-determined diseases each year in the United States alone.

Two inevitable questions arise. How large is the burden of cancer, leukemia, and genetic diseases expected to be from unbridled atomic energy development? And why do we go ahead with such atomic energy development at such a high cost to the health of this and future generations? Much of this book is devoted to answering both of these questions and to suggesting why the intelligent public must become informed and aroused to action.

Number of additional cases likely to be large

Our federal government, acting upon inadequate information, had specified how much radiation the average citizen may legally receive from atomic energy programs. We have estimated that this *legally* permitted radiation dosage would ultimately result in the following tolls:

- 32,000 extra cancer plus leukemia deaths annually for the current population of 200 million people.
- 150,000 to 1,500,000 extra deaths from genetically determined diseases annually for a future population of 300 million people. (This does not even include the genetically-determined stillbirths and infant deaths).

Recently a totally independent evaluation of the genetic hazard of radiation has been published by the Nobel Laureate, Professor Joshua Lederberg.* His estimate is that the Atomic Energy Commission's standards of permissible exposure would increase the natural rate of genetic mutation about 10%. Over a period of generations, the health cost of these additional mutations would be about $10 billion a year. Professor Lederberg stated further that uncertainties of the precise magnitude of the effect could mean an annual health cost of $1 billion to $100 billion per year.[1] This latter figure is in the neighborhood of the entire federal budget!

Professor Lederberg's estimate of a 10% increase in mutations from AEC permitted radiation dosages to the public is in excellent accord with our own estimate of between 5 and 50% increase in mutations, an estimate which led us to predict 150,000 to 1,-500,000 extra genetically-determined deaths annually.

The incredulous reader will, of course, ask why we could, as a society, be so foolish as to *permit* radiation dosages with such ultimate consequences. John F. Kennedy, in his book, *Strategy of Peace,* expressed his concern over radiation dosages from the fallout of weapons testing—dosages approximately *twenty times lower* than those now legally permitted for atomic energy development. These were his words:

> While many competent scientists agree that there has been no great harm done to mankind as a whole from the amount of radiation created by bomb tests so far, it is also true that there is no amount of radiation so small that it has no ill effects at all upon anybody. There is actually no such thing as a minimum permis-

---

*Joshua Lederberg is Professor of Genetics at the Stanford University School of Medicine, Palo Alto, Calif. He was awarded the Nobel Prize in Medicine for his contributions to genetics.

sible dose. Perhaps we are talking about only a very small number of individual tragedies—the number of atomic age children with cancer, the new victims of leukemia, the damage to skin tissues here and reproductive systems there—perhaps these are too small to measure with statistics. But they, nevertheless, loom very large indeed in human and moral terms. Moreover, there is still much that we do not know—and too often in the past we have minimized these perils and shrugged aside these dangers, only to find that the estimates were faulty and the real dangers were worse than we knew.

President Kennedy's compassion is matched by the profoundly prophetic features of the last sentence of this quotation. The estimates were *indeed* faulty and the real dangers are *much* worse than previously realized. Worse yet, we are currently developing several atomic energy programs while permitting a *twenty times* higher radiation dosage to our population than that which worried President Kennedy.

Atomic energy proponents spend a large fraction of their time and energies worrying that the public will be frightened by knowing the true health hazards of radiation. They have, therefore, mounted a public relations campaign of major magnitude to reassure the citizens that atomic energy programs are not delivering the legally permitted radiation dosage—yet. But they resist every effort to reduce the legally permitted radiation dosage. It is even more scary to observe this resistance than to contemplate the health hazard of the radiation itself.

Why do we witness this bizarre spectacle on the part of atomic energy technological proponents? It is the substance of this book to answer this question.

Many readers, having seen the magnitude of the health hazards described above, may prefer, for now, to pass over Chapter 2, which describes in simple, but quantitative, terms how the true hazard of radiation was grossly underestimated until recently. However, if we are not to prepare the way for rats, as Russell feared we may, it is hoped that the reader will sooner or later take the time to read Chapter 2. The technical details there will prove more easily understandable to the intelligent layman with an open, curious mind than to the "expert" who has long ago ceased to learn. The lay public has listened before to the findings arising in the sciences that deal with man; the lay public has understood, and it has acted constructively. To be sure, the scientific proof of the germ basis for many diseases was difficult. The public understands this scientific discovery today. It must not be forgotten that public understanding and support led to sanitation programs and immunization programs, both of which have contributed immeasurably more to eradication of germ-caused diseases than all antibiotics combined.

The scientific details concerning radiation injury to humans are no more difficult to understand than the germ basis of disease, but radiation injury is potentially far more destructive of present and future health than are germs. You can skip Chapter 2 for now, with no loss of understanding. But the lessons there of errors in public health aspects of technology are general lessons, important to appreciate and, we believe, worthy of your study.

2 Biological effects of radiation

A material introduced into our environment may represent a serious biological hazard for two reasons: It may represent a hazard (1) because of what we know, and (2) because of what we don't know about the material. Our knowledge can guide us, but we must be constrained by our ignorance. For essentially every material introduced into our environment, our ignorance pales our knowledge.

We know more about the biological effects of radiation than any other environmental pollutant. What we do know tells us that the present permissible radiation pollution levels are a travesty of public health. What we don't know concerning the genetic effects of radiation suggests a monumental tragedy. At long last we should heed Claude Bernard, who wrote, "In ignorance, refrain."

Radiation and the induction of cancer and leukemia

There exists by now a large body of detailed information concerning the injurious (including lethal) effects of ionizing radiations (alpha particles, beta particles, neutrons, x-rays, and gamma rays) on numerous animal species studied in the laboratory, as well as upon man. Indeed, the experimental animal data have long provided sufficient evidence concerning radiation hazards to have led to a rational approach for consideration of hazard to man. Unfortunately, and because of the promotional aspects of atomic energy development, the sanguine requirement has been made to see the human corpses before any credibility is assigned to the hazard for homo sapiens. In no other field do we proceed in such an incredible manner. In the food additive field, human use is barred if cancer production is demonstrated experimentally in *any* species. This represents elemental common sense as a public health approach. Such common sense has at no time characterized the approach in the field of radiation.

Thus a massive deception has been foisted upon the unwitting public in assessing radiation hazard for man. With respect to causation of leukemia and cancer in man the story is truly appalling. Those persons promoting this technology have demanded, and still demand, in evaluating hazards for man, that direct observation of corpses from each and every type of leukemia and cancer is required. This led to a colossal underestimation of the radiation hazard for man, and *still does,* in a manner important to describe here.

The Federal Radiation Council (FRC) guidelines for "Permissible" radiation

We measure the amount of radiation received by the whole body or any of its parts in the unit known as the *rad*. One rad is defined as the absorption of a specified amount of ionizing radiation energy (100 ergs of energy per gram of body tissue). An average individual receives 1.5 rads of radiation from a variety of medical procedures by 30 years of age. For most practical purposes the rad is equivalent to another unit commonly encountered and known as the rem. One rad equals one thousand millirads. Beyond these simple definitions of radiation dosage the reader will require no additional concepts to understand how radiation units are used in consideration of radiation-causation of cancer, leukemia, and other diseases.

There has been ample reason for skepticism concerning our Federal Radiation Council Guidelines which legally permit the average U.S. citizen to receive 0.17 rads (or 170 millirads) per year as a result of "peaceful" atomic energy activities. And such skepticism has been merited for many years. In essence, this is the case because a valid scientific justification for the allowable dose of 0.17 rads of total body exposure to ionizing radiation has never been presented. The general vague statement is usually repeated that the risk to the population so exposed is *believed* to be small compared with the benefits to be derived from the orderly development of atomic energy for peaceful purposes.

Dr. Brian MacMahon, Professor of Epidemiology at Harvard, writing as recently as early 1968, stated:

> While a great deal more is known now than was known 20 years ago, it must be admitted that we still do not have most of the data that would be required for an informed judgment on the maximum limits of exposure advisable for individuals or populations.[2]

This is vastly different from the bland reassurances of the Federal Radiation Council (FRC) Guidelines. We find ourselves in general agreement with Professor MacMahon, except that we go further and feel the already-documented evidence amply justifies a drastic revision *downwards*—and NOW.

There is an even more hazardous situation associated with the vagueness of the justification for FRC Guidelines. This hazard has become apparent to us through extensive contact with people in radiation surveillance work, in the atomic energy industry, and in atomic energy laboratories. Widely prevalent is the notion that the existing standards have a wide margin of safety built in. Many such individuals refuse to believe that any responsible body would ever set a guideline dosage into the federal statutes without a wide margin of safety.

How is it possible that our current FRC Guidelines may have falsely lulled us into complacency? Let us trace the evidence and restrict our considerations to two major effects of radiation upon human beings in this generation, namely, cancer and leukemia; that is, effects upon those actually receiving the radiation. *Any conclusion* we draw concerning the hazard of the current radiation guidelines can only be amplified and buttressed by consideration of the additional burden of human misery associated with genetic defects, fetal deaths, and neo-natal deaths. The case against perpetuation of the existing FRC Guidelines is overwhelmingly strong just on the basis of the cancer-leukemia risk, without even considering the potentially much larger problem of effects upon future generations.

How did the complacency arise?

First of all, there once existed a very great paucity of data concerning the radiation dose versus effect relationship between radiation and cancer or leukemia induction in man. Steadily, however, during these past 20 years, parts of the story have come to light from a combination of several extremely important sources:

(a) Study of survivors of Hiroshima-Nagasaki by the Atomic Bomb Casualty Commission.

(b) Study of patients *treated* with radiation for non-malignant diseases earlier in life and subsequently developing cancer or leukemia.

(c) Study of children who commonly received irradiation to the neck area in one unfortunate era of American medicine.

(d) Study of the occurrence of lung cancer in uranium miners in the United States.

(e) Study of cancer and leukemia in children whose mothers had received irradiation (diagnostic) during the pregnancy.

(f) Study of tuberculosis patients who had received extensive fluoroscopic radiation.

As the early results started to come forth from the Atomic Bomb Casualty Commission, it was noted that leukemia might be appearing more frequently in those persons irradiated in Hiroshima and Nagasaki. Attention became centered upon leukemia as a sort of "special" response to ionizing radiation and not much thought was given to other forms of cancer. From the Atomic Bomb Casualty Commission studies and from wholly independent observations, it is now clear and we believe no one disputes the estimates, that at least for total doses of 100 rads or more to adults, the leukemia risk may be expressed as follows:

One to 2 cases of leukemia per million exposed persons per year where each of them has received 1 rad of total body exposure. This does not require 1 rad per year; rather, we are talking about the above rate of disease occurrence with a total accumulated exposure of 1 rad. Furthermore, this incidence of 1 to 2 cases per million people per year persists for many years, once the latency period is over, ultimately declining somewhat, at least for chronic leukemia. It is a known fact, from many observations, that leukemia or cancer is *not* an immediate response to radiation. There is a period of years (different for different forms of cancer) before the clinical disease is manifest. This period is called the latency period. What happens during this latency period remains unknown. We simply know such a period exists.

An incidence rate of 1 or 2 cases per million people per year sounds like a small number, especially when this number is viewed in isolation. Indeed, many have hastened to add that spontaneously, without any man-made radiation, leukemia occurs with a frequency of 60 cases per million per year (varying with age) which makes it a relatively rare disease. So, 1 or 2 cases per year sounds small by itself, and sounds even smaller viewed against a spontaneous rate of 60 per million persons per year. And, as a result, with the early atomic bomb survivor data only showing leukemia, a widespread complacency set in concerning long-term effects of ionizing radiation, a complacency extending to high circles.

For two very major reasons, this error in thinking has turned out to be a mistake of the first order of magnitude.

First, leukemia happens to show a shorter latency period than most other forms of cancer. Therefore, the reason it appeared early to be the only malignancy in the Hiroshima-Nagasaki survivors was simply that not enough time had elapsed for the other forms of cancer to manifest themselves.

Second, the *proper* way to look at the incidence rate of 1 to 2 per million persons per year from radiation and the 60 per million persons per year spontaneously is not in isolation from each other, *but in relation* to each other. Thus, viewed in this light, 1 rad of ionizing radiation increases the leukemia incidence between 1.6 and 3.3%. Or, we can state that the doubling dose for leukemia (namely, that amount of radiation which will double the spontaneous rate) is between 30 and 60 rads. (Doubling a spontaneous rate of 60 cases per million each year means producing an additional 60 cases per million per year).

What about other forms of cancer?

It now becomes an issue of paramount importance to know whether other forms of cancer behave similarly in response to ionizing radiation. Are other forms of cancer describable by a fractional increase in occurrence rate per rad, and if so, how do the fractions compare with those for leukemia? We need no longer speculate about such matters because *hard, incontrovertible data* are available for human cancers induced by radiation. These data represent *facts, not opinion*. Estimates are available for several forms of cancer from worldwide data, United States data, and from the studies by the Atomic Bomb Casualty Commission of survivors of Hiroshima and Nagasaki. Let us consider a variety of forms of human cancers.

The extended study of the Japanese survivors of Hiroshima and Nagasaki provided clear evidence that excess cases of thyroid gland cancer, breast cancer, and lung cancer were occurring due to the radiation received, in addition to the leukemias which had already been recognized at an earlier time.

From the important studies on 14,000 human beings who received therapeutic radiation for the arthritis-like disorder known as rheumatoid spondylitis, Court-Brown and Doll* have discovered the subsequent occurrence of many forms of cancer in organs heavily exposed, incidental to irradiation of the primary disease in the spine.[3] And similar to the fate of the Japanese atom-bomb survivors, leukemia was the first form of cancer to appear in these patients. But beginning some ten years after their irradiation, many other forms of cancers began to appear

---

*The late W. M. Court-Brown was director of the Medical Research Council's Clinical Effects of Radiation Unit, Western General Hospital, Edinburgh, Scotland. Dr. Richard Doll is director of the Medical Research Council's Statistical Research Unit, University College Hospital Medical School, London.

in abundance. It is now clear that, in addition to leukemia, cancer has been caused by radiation in the following organs: lung, stomach, lymphatic and blood-forming organs, pancreas, pharynx, bone, colon, plus a variety of additional cancers of miscellaneous origin.

Last year we were endeavoring to make some reasonable sense out of the steadily accumulating evidence of additional forms of cancer being added as those proven to be produced by radiation of humans. We came to the realization that almost all the major forms of human cancer were by then already known to be produced by ionizing radiation. Radiation causation had already been proved for those types of cancer which make up 90% of the cancer mortality in the United States. For all intents and purposes it would hardly matter whether the remaining rarer, or minor forms of cancer could also be produced by radiation. So it became possible to state a primary principle, or "Law" of radiation production of cancer in humans.

That principle or law states, "All forms of human cancer are, in all probability, induced by ionizing radiation."

Following this important generalization, the very next question which arises concerns the quantitative aspects of the problem.

Our question can be restated now as, "How many extra cancers or leukemias will be produced per rad of radiation delivered to human beings?" Examination of the data accumulated from several sources demonstrated that a particular dose of ionizing radiation increased all cancers and leukemias approximately in proportion to the spontaneous incidence of those particular cancers or leukemias. By "spontaneous" is here meant those cancers and leukemias occurring without the insult of man-made radiation. This important generalization, that radiation produces cancer and leukemia in proportion to spontaneous occurrence rate, can best be appreciated by reference to specific cancers and leukemias.

In the following table is shown the evidence concerning cancer and leukemia production by radiation. The data are presented in two ways, (a) as the percent increase in cancer occurrence rate for each rad accumulated by the exposed persons and (b) as the number of rads of exposure that would result in producing a number of cancer cases equal to the number occurring spontaneously each year. This latter number is commonly referred to as the "doubling dose" of radiation for cancer or leukemia production. If one rad of radiation produces a 1% increase in occurrence rate for a particular cancer, then 100 rads is the doubling dose; if one rad produces a 2% increase in occurrence rate, then 50 rads is the doubling dose. One simply divides 100 by the percent increase to arrive at the doubling dose for any cancer.

TABLE I*

How Radiation Increases Occurrence Rate of Various Cancers and Leukemias (Adults)

Type of Cancer	Percent Increase in Occurrence Rate for Each Accumulated Rad	Number of Rads of Radiation Required to Double Spontaneous Cancer Occurrence Rate
Leukemia	2.2%	45 rads
Thyroid	1.0%	100 "
Lung	1.4%	70 "
Breast	3.3%	30 "
Stomach	1.1%	90 "
Pancreas	1.7%	60 "
Bone	1.1%	90 "
Lymphatic+Other Blood-Forming Organs	2.2%	45 "
Colon	1.1%	90 "
Miscellaneous Cancers	2.2%	45 "

For such an array of widely divergent human organ systems, already including very firmly established evidence for nearly all the major forms of human cancers, it is amazing indeed that there is such a small range for the estimated doubling dose. Correspondingly, there is a very small range in the estimated

*Prepared by the authors, this table is a summary of data from all sources and represents their best estimates of the doubling dose for human cancers.

percentage increase in occurrence rate per accumulated rad for these widely different organ sites in which cancers arise.

Cancers vary as to the latent period between the receiving of radiation and the clinical observation of the cancer. Many of the forms of cancer shown in the table have long latency periods. At a result, the true increase in cancer due to radiation will be even higher than shown in the table, for the full effect has not yet been seen in some of the human groups exposed to radiation. Therefore, we have concluded that the true average doubling dose for all forms of cancer produced by radiation will not be larger than 50 rads, and consequently for each rad of radiation there is not less than a 2% increase over the spontaneous occurrence rate for each form of cancer.

Thus, the second important generalization, or law, arrived at is the following: "All forms of cancer and leukemia are increased by ionizing radiation in direct proportion to the spontaneous occurrence of such cancers or leukemia. Best estimates at this time suggest a 2% annual increase in incidence rate for every form of human cancer and leukemia for every rad of radiation accumulated by human adults."

Whether the value of 2% increase will finally be the precisely correct one depends upon further data accumulation. What is of moment is that the increase is indeed very large.

Previous estimates of radiation hazards were too low

Some forms of cancer are spontaneously much more rare than others. This "law," or generalization, states that both the common and the rare forms of cancer are increased about 2% in occurrence rate for each accumulated rad. Early attention of radiation scientists focussed on leukemia, which is, fortunately, relatively rare. But we must now realize that all other forms of cancer combined occur approximately *twenty times* as frequently as leukemia. Thus, if one rad of total body radiation will produce two cases of leukemia per year in a particular group of human beings, that same radiation dose will produce 20 x 2, or 40 extra cases of all forms of cancers per year.

This is the real import of the second generalization above, and this is the reason we ourselves were so shocked by the implication of our observations that had led to this generalization. The radiation cancer hazard was thereby shown to be huge in contrast to the previous estimates (20-fold higher than leukemia alone)—estimates that were all falsely low simply because leukemia became manifest in radiation-exposed humans earlier after the radiation than did the many diverse forms of cancer.

This generalization leads directly and simply into the estimation of the devastating effect upon human life if the current FRC Guidelines for population exposure are allowed to remain in force. These Guidelines have never had any more than a "by guess and by gosh" justification. About all that ever was claimed even by the Federal Radiation Council itself was the hope that somehow the benefits to be derived from allowing human beings to be radiated to this extent would outweigh the harm that could accrue. That harm, from cancer and leukemia alone, is easily estimated from the generalizations below:

(a) 1 rad increases all forms of cancer in adults by 2%.

(b) Federally allowable dose accumulation is 0.17 rads per year, or 5 rads from birth to 30 years.

(c) 5 x 2 = 10. So, accumulating 5 rads leads to a 10% increase in cancer + leukemia occurrence.

(d) Since approximately 320,000 cancers occur spontaneously each year in the USA, a 10% increase means some *32,000 extra cancers plus leukemias each year in the USA*.

In examining the available data on radiation-induction of human cancer, additional profoundly important and shocking evidence was found. Two supremely important groups of humans provided this evidence: irradiated children and fetuses irradiated in utero. For the children irradiated during infancy in the region of the thyroid gland, thyroid cancer was later found to be caused by this irradiation. Instead of 50 rads being required to double the spontaneous cancer incidence (as in adults), it turned out that only approximately 5-10 rads were required. This indicated a strikingly greater sensitivity of children in susceptibility to radiation injury.

For infants in utero the situation is even much worse. From the great on-going work of Stewart and Kneale,* important new evidence recently published shows that all forms of childhood cancer and leukemia are doubled by *extremely* small dosages of radiation.[4] If this radiation is delivered in the first 13 weeks of pregnancy, only one-third of a rad is sufficient to double the occurrence of childhood cancers and leukemias. Later in pregnancy, approximately 1½ rads causes such a doubled rate of occurrence of cancer and leukemia. All of this evidence indicates exquisite sensitivity of the infant in utero, even compared to already high sensitivity of children to radiation injury. Moreover, these important studies of Stewart demonstrate the validity of the assumption of the International Commission of Radiological Protection that any amount of radiation, no matter how small, is harmful to human beings.

It is not at all surprising that infants in utero should appear most sensitive to irradiation, children next in sensitivity, and adults third (but *by no means low*). This is precisely the order in which these groups stand in terms of the fraction of their cells undergoing cell division at any time—and much evidence suggests these are the cells most susceptible to cancer induction.

General laws of cancer induction by radiation

We can now summarize the general laws for cancer production by radiation of human beings, including the evidence for infants in utero, children, and adults.

Law 1 "All forms of cancer, in all probability, can be increased by ionizing radiation, and the *correct* way to describe the phenomenon is either in terms of the dose required to double the spontaneous incidence rate of each cancer or, alternatively, as the increase in incidence of such cancers per rad of exposure."

---

*Dr. Alice Stewart is Chairman, and G. W. Kneale is Henry Goodger Scholar, Department of Social Medicine, University of Oxford.

Law II "All forms of cancer show closely similar doubling doses and closely similar increases in incidence rate per rad."

Law III "Youthful subjects require less radiation to increase the incidence rate by a specified fraction than do adults."

Based upon these laws and the extensive data already in hand and described above, the following assignments appear reasonable for all forms of cancer and leukemia:

For adults 50 rads to double the spontaneous cancer and leukemia incidence. 2% increase in incidence rate per rad of exposure.

For children Between 5 and 10 rads to double spontaneous incidence. 10 to 20% increase in incidence rate per rad of exposure.

For infants in utero Between ⅓ and 1½ rads to double spontaneous incidence. 60 to 300% increase in incidence rate per rad of exposure.

For the radiation of infants in utero, Stewart and Kneale had clearly stated the outlines of these general laws. For adults, Court-Brown and Doll had clearly stated the outlines of these general laws.

With all the additional data available, plus the data of Stewart and Kneale, MacMahon, and Court-Brown and Doll, we consider the enunciation of these general fundamental laws as having a better experimental base than many laws of physics, chemistry, or biology had when first proposed. Furthermore, we would estimate that the absolute numbers, if anything, probably underestimate the risk. For purposes of setting radiation tolerance guidelines, one might even be advised to use lower doubling doses than those estimated above.

The implications of these laws for the population exposure associated with Atoms-for-Peace programs

The statutory allowable dose to the population-at-large in the USA is 0.17 rads per year from peaceful uses of atomic energy in all forms. If everyone in the population were to receive

0.17 rads per year from birth to age 30 years, the integrated exposure (above background) would be 5 rads per person. If the risk for all forms of cancer plus leukemia is an increase of 2% in incidence rate per rad, we have 5 x 2 = 10% increase in incidence rate for all forms of cancer plus leukemia per year for a population of two hundred million persons in the USA. And this estimate does not even credit the much greater sensitivity to cancer induction of the radiated child or infant in utero. Our estimate of 32,000 extra cancers and leukemias is probably too low as a result of not taking the child's sensitivity into account. And 32,000 extra cancer plus leukemia cases per year exceeds by far the mortality rate from the highest mortality year in the Vietnam war!

It seems to us that this alone is rather a high price to consider as being compatible with the benefits to be derived from the "orderly development of atomic energy."

And we must add to these death estimates the comment that we have used only the hard data in hand based upon cancer and leukemia induced in humans by radiation. We have said nothing thus far of the additional burden of loss of life and misery from genetic disorders in future generations, fetal deaths, and neo-natal deaths. Furthermore, we have not used the vast array of experimental animal data which indicate that not only does cancer mortality increase from irradiation, but that many, if not all, causes of death increase—and in about the same proportion as does cancer mortality. In human beings this could multiply by four times the 32,000 extra cancer plus leukemia deaths per year in the United States.

What must be done

It would appear that the only sensible thing to do right now is to reduce *drastically* the Federal Radiation Council radiation dosage guidelines—guidelines which permit such excessive radiation dosages to the population. We draw no comfort from the fact that everyone in the population is not yet receiving this "allowable" dosage. With a variety of burgeoning atomic energy programs, an increasing proportion of the population will indeed receive increasing dosages of radiation. Delivery of even a small fraction of the dosage currently legal would be an unspeakable tragedy in view of the absence of any justification for such an irresponsible act.

This, in essence, is the evidence and the message we delivered before an eminent scientific society, the Institute of Electrical and Electronic Engineers, on October 29, 1969, in San Francisco, California. And this was the message we repeated on November 18, 1969, to the Senate Subcommittee on Air and Water Pollution which was holding hearings on matters related to the hazards of radiation from various atomic energy programs. In those presentations we urged the Atomic Energy Commission to join us in the effort to achieve a much safer set of Federal Radiation Guidelines.

The Atomic Energy Commission did not join us. Instead the staff of the AEC criticized *where* we had presented our findings, *when* we had presented our findings, and *to whom* we had presented our findings. None of these criticisms dealt even remotely with the issue at hand—namely, a grave hazard to the public health. We felt it was urgent for everyone to know whether the Atomic Energy Commission had any desire to get at the truth of our findings. We were anxious for the best scientists in the country to go over our scientific evidence with a fine-tooth comb, to criticize our evidence, and to ask us any questions they wished.

On January 28, 1970, we were presenting the evidence described above, plus a great deal of additional confirmatory evidence based upon further study, at hearings of the Joint Committee on Atomic Energy. In the halls of Congress on that date we issued the following invitation to the Atomic Energy Commission:

A Scientific Challenge to the Atomic Energy Commission Staff Concerning the Cancer + Leukemia Risk from Radiation

Chairman Holifield, we urge you to nominate a jury of eminent persons, physicists, chemists, biologists, physicians,

Nobel Prize winners, or National Academy of Science members, or American Association for Advancement of Science members —none of whom have any atomic energy axe to grind. We urge you to serve as chairman of a debate. Dr. Tamplin and I will debate each and every facet of the evidence concerning the serious hazard of Federal Radiation Council Guidelines against the entire AEC staff plus anyone they can get from their 19-odd laboratories, singly, serially, or in any combination.

With their 20 year background on this problem and their large staff to draw on they should be razor-sharp at a moment's notice. We are ready now. If there is any valid reason for questioning our submission to peers and for questioning our evidence, this eminent jury of peers will certainly determine so. If the debate before eminent peers is not held, then by default, we think the entire country and the world will know the answer without further question.

That was January 28, 1970.
Now it is the latter part of 1970.
The AEC Staff has not been heard from.
It appears as if the true answer is known by AEC default.

Genetic consequences of radiation

It is important to point out at the onset of any discussion of the genetic effects of radiation that we know the effects are harmful but we do not know precisely how harmful. This is emphasized by the United Nations Scientific Committee on the Effects of Atomic Radiation:

> Since neither a comprehensive estimate of the genetic risk, nor an upper limit to that estimate is available, the assessment of genetic damage from main sources of radiation must still be made by means of comparative risks.[5]

In other words, we know that 10 rads of radiation are worse than 1 rad, but we don't know how harmful 1 rad is. At the same time, in the United States and elsewhere, we have a so-called "permissible" level of exposure. This level corresponds to that recommended by the International Commission on Radiological Protection. The public is often told that this level is safe or negligible. But consider the highly qualified aspects of this recommendation:

> Because of the need for guidance in this regard, the Commission in its 1958 Recommendations suggested a provisional limit of 5 rems* per generation for the genetic dose to the whole population, from all sources additional to natural background radiation and to medical exposures. The Commission believes that this level provides reasonable latitude for the expansion of atomic energy programs in the foreseeable future. It should be emphasized that the limit may not in fact represent a proper balance between possible harm and probable benefit, because of the uncertainty in assessing the risks and the benefits that would justify the exposure.[6]

Notice that they indicate that a major consideration was allowing a "reasonable latitude for the expansion of atomic energy programs." One wonders whether the recommendation of this small group of "experts" should be accepted without question. Is the "reasonable latitude" really reasonable? Should not a much broader segment of society than this small group make this decision—a decision for all men and all time—where man's heredity itself is at stake?

There are two components to the genetic effects of radiation. (1) Lethal effects that lead to death before maturity or that lead to sterility, and (2) effects that contribute to the general pattern of illness and mortality in adult life. In the population the present pattern of illness and mortality results from a complicated (and essentially totally unknown) interplay between heredity factors and the environment. The mechanism by which radiation would be expected to influence these patterns is by altering the genes and chromosomes that determine the heredity factors transferred to the child.

The United Nations Scientific Committee on the Effects of Atomic Radiation states:

> It is generally accepted that there is a genetic component in much, if not all, illness. This component is frequently too small to be detected; in other instances the evidence for its presence

---
*1 rad is equal to 1 rem for most of the radiations encountered.

is unequivocal. Nevertheless, the role of genetic factors in the health of human populations has not in the past been considered seriously in vital and health statistics. As a consequence, data on the prevalence of hereditary diseases and defects are now largely restricted to that collected by geneticists for special purposes in limited populations from a small number of countries.

An assessment of the hereditary defects and diseases with which a population is afflicted does not necessarily provide a measure of the imposed burden of suffering and hardship on the individual, the family, or society.[7]

This can be paraphrased as "most of our information concerning genetic disorders in man relates to simple gene mutations in such rare diseases as hemophilia. This is only the top of the iceberg because *all* human disease has a genetic component."

The United Nations Committee goes on to state that most human disease has a genetic component but it is not related in a simple way to a simple dominant-recessive gene system. The major human diseases are determined by the interplay of a large number of genes and the environment in an unknown fashion.

Just recently Dr. C. O. Carter* published conclusive evidence demonstrating a multi-gene basis for such major diseases as diabetes, ischemic heart disease, schizophrenia, and rheumatoid arthritis.[8] Ischemic heart disease (coronary heart attacks) kills two or more times as many Americans annually as all forms of cancer combined. The toll of schizophrenia socially is best stated as massive.

In estimating the genetic effects of radiation the above considerations indicate that it is essential to assume that all of man's illness and mortality are a consequence of mutations in the population. By this assumption, if the mutation frequency were doubled, these death and disease rates would be doubled. This is the only reasonable assumption.

The United Nations Scientific Committee on the Effects of Atomic Radiation estimates that 1 rad would increase the

---

*Dr. C. O. Carter is Medical Genetics Director, Clinical Genetics Research Unit, Institute of Child Health, London.

natural mutation frequency by a factor between 1/10 and 1/100. The existing radiation protection guidelines would allow a genetically significant dosage of 5 rads to be accumulated by 30 years of age. This could increase the mutation frequency and hence increase the death and disease rates between 5% and 50%. As an upper limit then, the radiation protection guideline dosage could increase the death and disease rates by 50%. It is difficult to understand how the present allowable exposure of 5 rads in 30 years is justified even if the true effect were the smaller value; i.e., 5%.

Earlier optimistic reports were erroneous

As we indicated in the earlier sections of this discussion, the more recent data on the biological effects of radiation are generally tending to demonstrate that the original optimistic opinions of the effects were wrong. For example, we now realize the extreme radiosensitivity of the developing fetus in utero. To a considerable extent the existing guidelines were based upon the effects of radiation on adults, and the data of Stewart and others are now demonstrating that the developing fetus is from 25 to 150 times more sensitive than the adult. In addition, the data are suggesting that leukemia is not the most sensitive form of cancer with respect to radiation, but that indeed all cancers which occur in the population more frequently than leukemia are induced by radiation in proportion to their occurrence rate.

The early estimates of the genetic effects of radiation were based upon studies of the fruit fly (Drosophila). Unfortunately, the extension of the genetic studies to the mammal (mouse) revealed that the genetic mutation rate per unit dose of radiation was much higher than that observed in Drosophila. This should certainly be a cause for caution in extrapolation of the mouse genetic data to humans.

Finally, the data which are now coming in from biological experimentation are suggesting that the most radiosensitive portion of the biological system was overlooked in setting the original standards. This part of the biological system is represented by the chromosomes; that is, the "packages" of genes.

Radiation can affect genetic material in two major ways. One way, which is the one that has been given the most attention with respect to the genetic and somatic effects of radiation, is that of producing a point mutation. By this process irradiation changes the structure or the composition of a single gene; that is, a single hereditary unit. Through this process is determined the production of a point mutation.

Damaging effect of radiation on chromosomes

On the other hand, it is now abundantly clear that the radiation can also affect the chromosomes. By this process the radiation is able to alter or remove from the genetic material not a single gene but a large number of genes. The developing evidence on chomosomes and the effects of radiation on chromosomes suggests that this process may represent the major mechanism through which radiation produces its damage, such as cancer induction in the irradiated person and many forms of disease and debilities in his offspring and future generations. In other words, a whole new mechanism for the potential biological effects of radiation is now evolving, a mechanism that may represent a far greater susceptibility of man than any previous mechanism proposed. What should be shocking is that only now are we beginning to learn about the massive disease and death producing effects of alterations of chromosomes—a body of knowledge beginning to be accumulated after the experts, totally ignorant of such phenomena, had already decided on so-called "acceptable" doses of radiation.

And what does the U.N. Committee have to say about this new body of evidence?

> Present knowledge of dosage effects on the induction of chromosome anomalies is too scanty to predict a doubling dose. There are indications that monkey chromosomes and, hence, perhaps those of other primates are more radiosensitive than those of mice. The Committee is of the opinion that ionizing radiation would increase the prevalence of developmental con-

genital malformations and of serious constitutional disorders, but no quantitative estimates can now be made.[9]

We think it is extremely important at this juncture to point out that scientists are not omniscient. Though we have a considerable body of information at our disposal, we can never be certain that we have made all the pertinent observations that are necessary to determine the outcome of a particular series of events. We must always keep in mind that we do not necessarily have all the significant facts before us when we are asked to make recommendations as to whether something which is planned will not adversely affect man or his environment. The recent developments that occurred with thalidomide represent a useful example in this respect. As a consequence of the rather tragic results of the use of thalidomide, the drug testing procedures have now been altered.

Following the thalidomide disclosure, we now find many drugs listed in the Physicians Desk Reference that have a pregnancy warning which was not recorded previously. In fact, as ridiculous as it may seem, drugs in the Physicians Desk Reference that were initially issued primarily for the treatment of nausea in early pregnancy have a warning against their use in early pregnancy.

It would seem that we have a similar situation now with respect to the biological effects of radiation. Subsequent to the establishment of the radiation exposure guidelines, a whole new body of experimental data concerning the radiosensitivity of chromosomes has been evolving. Recent results reported by a group from Johns Hopkins University may demonstrate quite well the importance of the new body of data with respect to the biological effects of radiation. The Johns Hopkins data indicate that between 1 and 2 rads delivered in the first 30 weeks of in utero life will produce severe genetic damage, and in this case it appears to be chromosomal damage to the fetal germ cells. As a result of this damage, 50% of the female conceptuses of women who were themselves irradiated as fetuses will be killed.[10] This is a very startling observation, and other confirmations of this observation are necessary and highly desirable. It is an effect the magnitude of which far exceeds anything that had previously been predicted concerning the genetic effects of radiation.

How serious are the genetic effects of radiation? No one knows! It is possible that exposure to the present allowable levels could result in a 5% to 50% increase in the death rate, producing some 150,000 to 1,500,000 additional deaths each year in a future population of 300,000,000 people. Moreover, the evidence suggests that there would be (over and above the fatal diseases) a 5% to 50% increase in such crippling diseases as diabetes, rheumatoid arthritis, and schizophrenia. With the present radiation guidelines we will be practicing an insidious form of sadism and genocide.

3 Beware the Gallant Knight of Technological Progress

The presumption is widely made that in some mysterious way a rational approach will be forthcoming that will enable man and his ecosystem to survive environmental degradation. With much fanfare, a stern glance is cast upon a particular smokestack, the latrine output at West Point, or an oil slick in the Gulf of Mexico. As a result of the fanfare, it is assumed that a new day of attention to our environmental integrity has dawned when, in fact, all the irrational approaches of the past are tenaciously defended and all suggestions of a rational future approach are slandered and ridiculed as the work of conservationist "kooks."

For those who do not understand that our environmental crisis is other than accidental, it is essential to review the nature of technological application of scientific discovery. It is within this general approach to development of technology that all the difficulties lie and the future catastrophes are carefully prepared. Nothing is left to chance. Our approach of the past and, unfortunately, of the present virtually insures disastrous ultimate results.

So similar are the various technologies in this generation of their respective parts of the environmental crisis that almost any of them could be chosen for illustrative purposes. All such technologies can be regarded as "polluters" of man himself, of his supporting ecosystems, or of the inanimate environment.

Radiation exposure and radioactive contamination of the biosphere make up the problem of radiation pollution. An argument could be made for radiation pollution as deserving the Number 1 position among environmental hazards; as good an argument can be made that other approaches to extermination of life on this planet preempt this leading position because we won't be around long enough to suffer from radiation destruction. What is important about radiation pollution is that its study exemplifies all the errors of the past and the formidable obstacles to reasonable action in the future. The history of the automobile would serve as well, but since our area is atomic energy, we shall attempt to outline the general principles which concern us by using radiation pollution as a framework of specifics.

THE ATOMIC ENERGY COMMISSION: A CASE STUDY

The elements of the problem can be discussed under the following categories: (1) A wondrous New Technology is born, (2) A Gallant Knight champions the New Technology, (3) The Technology has a by-product that is a hazard to man, (4) The concept of a "tolerance" dose of radioactive poison is promulgated, and (5) The conflict between the Gallant Knight and his experts, and the public.

A wondrous New Technology is born

Roentgen's discovery of the X-Ray in 1895 and Becquerel's discovery of natural radioactivity in 1896 are key points in the field of radiation pollution. Any attempt to gainsay the remarkable nature of the phenomena involved would be foolish. Clearly these discoveries brought a new dimension to chemistry, physics, and biology—a dimension best described in two ways:

1. The packet of energy involved is massive compared with the previously familiar infra-red, visible, and ultra-violet packets of light energy.

2. Atoms, or more precisely nuclei, undergo transformation where the energy release per transformation is as much as a millionfold or more that release accompanying chemical transformations, e.g., as in the oxidation of carbon to carbon dioxide.

Man's curiosity and his quest for an ostensibly better life inevitably lead to a rapid exploration of the possibilities for exploitation of the new discoveries—in medicine, in industry, in warfare, and in the furtherance of scientific investigation itself. It would be difficult indeed to find a scientific discovery or technical development where this sequence of events does not occur. Roentgen's x-rays and the Curies' radium were very rapidly introduced into medical diagnosis and medical therapy. While, in retrospect, one may look with horror at the rashness of man in exploring his new technology, history teaches us that such rashness has not shown any signs of abating. In medicine, the frustration over the inability to cope with unsolved major diseases at any point in time is understandable, even if it leaves one shaking his head at the readiness with which almost anything new is tried.

Here we see the introduction of the cult of worship of technical progress, of the scientific method, with the undying confidence that science and technology are progress; more progress is up, and up is (by definition) good.

There is always the assurance that the new technology must be wonderful, that any appearance otherwise is but an indication that we have not yet learned how best to extract the full measure of wonders the technology or new scientific discovery has to offer. And even before this is learned, if indeed it be there to be learned, the "good is up" philosophy leads to a rapid expansion of the science and its technology.

Huge energy supply supports this New Technology

In the field associated with radiation and radioactivity, some of the important mileposts are the discovery of the nucleus of the atom, the neutron, artificial radioactivity, nuclear fission, and the self-sustaining nuclear fission chain reaction. The net result of all these epochal advances is that nuclear energy release is possible in fantastic quantities, and radioactive substances are available in massive abundance, packagable with a richness of varieties that put the famous Heinz-57 to shame.

Since "good is up" in this field, it is believed that the "good" must indeed be super-marvelous, for the "up" is certainly astronomical.

In the early stages of an area of science or science plus its technological offshoot, the practitioners are not many and, rash as they may be, their small numbers preclude, in general, global consequences of stupidity and poor judgment. But, as with the technology itself, the practitioners themselves believe progress means more practitioners, more centers dedicated to the particular science and technology, more applications of the technology. This inevitably breeds more "progress" in the technology, the requirement for more practitioners, etc. in an ever-upward spiral.

Somewhere in this chain of developments, the science or technology becomes consequential enough to require a public relations agency, a lobby for the activities. For small science or small technology, the professional associations and the business associations subserve these functions. In the field of radiation, radioactivity, and atomic energy, the phenomenon comes to require the same Madison Avenue approach applied by Detroit with the automobile. The time for the Gallant Knight Champion has arrived; it is too late for the small professional or business association.

A Gallant Knight champions the New Technology

The Gallant Knight approach can arise in either the private or the public sector, depending upon circumstances; the final result is the same. In either case a super-agency is the fundamental requirement.

For the automobile, the super-agency was achieved in the form of "Detroit," a triumvirate of corporations, aided by Madison Avenue, and dedicated to bringing the supreme benefits of the automobile way of life to every hamlet in the land. So exquisite was this new way of life that the ultimate goal of success in life became the achievement of the two-car garage, the total obliteration of the landscape by the freeway, and the final denial even of maintaining the physiological integrity of man (and his family) in the absolute requirement to devote his resources to the care and feeding of the automobile. Surely the super-agency, "Detroit," did its job of insuring the widest possible spread of this stunning technology unflinchingly and faithfully.

In the field of radiation and atomic energy, the peculiar circumstance that a military application itself led to major parts of the scientific-technological spiral has dictated the establishment of the requisite Gallant Knight super-agency in the public sector rather than in the private sector. This was achieved, by act of Congress, in the establishment of the Atomic Energy Commission, with the specific mission (in the non-military area) of bringing the benefits of the atom to the populace, with (what has become an afterthought) due consideration of safety and health of the public. When "Detroit" exercised its role of promoter of the automobile, the issue of public health and safety was hardly noticed (until recently, *very* late in the game).

Conflict between promotion and regulation

The question of regulation in relation to promotion hardly existed. In the establishment of the Atomic Energy Commission, due notice was given to the regulation part of the story because of the appreciated power of the technology itself, but one of the fundamental errors of history was made by the Congress in assigning to one super-agency both the role of promoter and regulator. Obvious as the conflict between the promotional and regulatory activities of the Atomic Energy Commission has become, there is vigorous denial by the AEC and by the super-promotional Joint Committee on Atomic Energy that any conflict in these roles exists. Unctuous statements of self-praise abound from both the AEC and JCAE, and sycophants are available within science and industry in abundance to confirm the mythology of a consistency in promotion and regulation. We shall return to this conflict later.

As a super-agency dedicated to the widest spread of the new technology, the AEC has truly rivalled Detroit, figuratively and literally, unfortunately. The AEC has several product lines, as has Detroit, and it is vigorous in merchandising all of them at once. Among these products are nuclear energy itself, available for conversion to heat and electricity, atomic and hydrogen bombs, radiation sources, and radioactive substances by the carload. One criterion alone signals successful execution of its mission for the AEC, namely, an ever-rising curve of output and distribution of all its product lines.

Nuclear electricity is heralded as cheap, even though its true expense is hidden by fantastic overt and covert tax-supported subsidiaries and the necessity to kill uranium miners with lung cancer to achieve such "cheapness." Nuclear electricity is "pollution-free" because its poisonous radioactive by-products are not optically visible as are the belching columns of smoke from poorly-designed fossil-fuel generating plants.

Nuclear bombs, unpopular when exploded directly upon the population, still need merchandising. According to the AEC, there are relatively few human needs that cannot be fulfilled with an appropriately designed nuclear explosive, atomic or hydrogen. Canals can be dug, harbors created, mountain cuts made, rivers diverted—all with nuclear explosives, provided one doesn't look too deeply at the residual radioactivity spewed and strewn about the earth. And we can, thanks to the foresight and hard work of the AEC promoter, have all the harbors, canals, etc. that we want, because nuclear bombs are now cheap! We are going to have simulated yields of underground natural gas, oil from shale, and metals from ore deposits—all by exploding underground nuclear bombs. Never mind the radioactivity associated with the gas, the oil, or the metals recovered. Indeed, recently a major promise for ridding the earth of garbage has come from promoters of nuclear explosives, via the technique of creating huge underground garbage pits by explosion of large nuclear bombs. And all these fabulous benefits are upon us—we can have thousands of such explosions per year in the immediate future.

With characteristic frugality, the AEC promises us it will stay economical and, like Armour's pig, use everything but the squeal. Its refuse by-products are being made available in the form of radiation sources of great intensity, and radioactive substances to assist every scientific, industrial, and medical endeavor. The curve of shipments of such radiation sources and radioactive substances rises annually, signalling great success in the frugal exploitation of by-products of the technology. That a probably large, and largely unknown, fraction of the radiation and the radioactivity finds its way by numerous routes into the biosphere, including man, in a cumulative fashion, is almost wholly overlooked, for the recipients and users are "licensed."

So, the Gallant Knight of atomic technology, the Atomic Energy Commission, fulfills its mission of champion of the technology with splendor, assisted by liberal use of tax dollars for massive public "education" (in the parlance known as propaganda).

The Technology has a by-product that is a hazard to man

The environmental crisis is upon us for a very simple reason: few technologies are free of side effects. Only recently have we come to realize that *many* of our technologies have side effects of such potential magnitude as to be capable of obliteration of massive segments, if not all, of our ecosystems. Atomic energy can be regarded as an archetype. Indeed, it is especially significant to consider radiation as a prime example because the lethal side effects were apparent shortly after Roentgen's discovery of x-rays. Yet in spite of this, there is not a shred of evidence that this long-standing knowledge has made the development of atomic energy technology one iota more rational than those with far more obscure side effects. How is it possible for the side effects to remain so poorly appreciated?

We must recall that technology is wondrous and that "good is up." Thoroughly imbued with these items of "knowledge," the super-agency gallant knights feel deeply their responsibility to reassure the public that we can have the exquisite benefits and solve the side effects problem by devoted research and development. Chairman Glenn T. Seaborg of the AEC is expert in the reassurance approach to coping with side effects. "We must learn to live with the atom wisely," he intones, "and this we are doing well." Other equally prominent atomic energy proponents tell us that "undue alarm can stifle progress" and "progress has brought us all the fruits of civilization we have." Professor Edward Teller* recently sought to allay the concern about radioactivity voiced by Senator Mike Gravel with the reassurance that our methods of getting radioactivity out of people are improving all the time. If we expose people to radioactivity, we'll clean them up, Professor Teller assures us.

At some stages in a technology, such as atomic energy, the platitudinous reassurance approach suffices to quiet public fears, especially if the platitudes are repeated on a regular schedule.

The difficulty for the technology arises, however, from the fact that the side effects are real, and they fail to melt away under platitudinous reassurances; that is, the side effects are obvious and cannot forever escape public awareness. Precisely this has happened in atomic energy, much to the consternation of the AEC, for it is truly a nuisance in the path of its mission to bring society the blissful benefits of the atom.

Side effects in 5 to 20 years

Thus, as a result of medical and industrial uses of radiation and atom bombings, a large number of humans have been exposed to radiation. If the side effects (as for example, cancer or leukemia) had been immediate, the AEC would have been seriously embarrassed many years ago. For reasons not yet understood, such side effects of radiation exposure as cancer or leukemia require 5 to 20 years to manifest themselves. Side effects such as irreversible damage to the genes will only appear in future generations of the individual radiated. This delay in manifestation of side effects has proved enormously useful to

*Dr. Edward M. Teller is Professor of Physics-at-large, University of California, Berkeley; also, Associate Director for Physics, University of California, Lawrence Radiation Laboratory, Livermore, Calif.

the AEC, for it has been able to carry on activities involving radiation of humans for many years, and each year point out that the humans are still alive (at least until the 5 to 20 years had elapsed and the leukemias and cancers became obvious to everyone).

The "tolerance" dose concept of radioactive poison is promulgated

The reader will certainly ask how the promoter, the AEC in this case, ever arrived at a certain radiation dose as "tolerable" or "permissible." This goes to the very heart of the problem all technology promoters face when side effects rear their ugly heads. The AEC, like other super-agency promoters of technology, has as a prime objective the growth ad infinitum of its technology. The first step is a total denial that side effects are related to the technology, especially where the individuals exposed to noxious by-products, such as radioactivity, don't drop over dead immediately upon exposure. Denial of these delayed effects buys the promoter time for unbridled exploitation and rape.

But a second line of defense is soon needed. This is the so-called "tolerance" or "allowable" dose approach used by the AEC and all other technology promoters. For most poisons, the fraction of people killed by the poison goes up with increasing dose. It becomes obvious, therefore, that lethal effects are perceived early in groups of persons who are massively exposed, say, to radioactivity as a poison. Suppose 100 persons are exposed to a particular dose of radioactivity and that 50 of them die of cancer. The promoter of atomic energy technology then comes up with an ingenious pronouncement—that particular dose of radioactivity must have been above the "tolerance" or "allowable" dose.

So, a new "tolerance" dose is set somewhere below this amount of radioactivity exposure, even though not one bit of evidence exists that *any* dose will be tolerated without producing cancer or leukemia. But for a promoter of technology, like the AEC, taking away the concept of a "tolerance" dose is far worse than taking candy from a child. For, so long as a "tolerance" or "allowable" dose can be re-set at some lower value, more time is available for the technology to proceed before it becomes obvious to everyone that even at the lower dose, grossly unacceptable numbers of humans are being killed by cancer and leukemia. It's just that it takes longer to prove that 5 out of 100 people are being killed than it takes to prove 50 out of 100 are being killed. The promoter thus buys more time for exploitation.

Actually, any poison that kills one extra human being out of 10,000 is a major public health disaster. So between the proof that 50 out of 100 are being killed by a radioactive poison and the proof that 1 out of 10,000 is being killed, there are innumerable opportunities to keep setting the "allowable" dose successively lower and, hence, buying more and more time. During all this, the atomic technology can flourish, tens of thousands of people can be murdered annually—all legally—under the deception entitled "within the allowable tolerance." And, as a dubious bonus, the hereditary gene pool of the human race is irreversibly damaged, to say nothing of irreversible contamination of the planet.

As the experience from human exposure to radiation and experimental animal exposure has accumulated, it has become painfully clear that no evidence exists, or has ever existed, that suggests there is any safe tolerance dose of radiation.

A rational approach is needed

How then did we ever get into this irrational box of the "tolerance" dose for a variety of technologically-produced poisons? We did so because of the role of super-agencies as promoters of technology. One doesn't have to consider the Atomic Energy Commission, which is a prime example, as a group of evil men dedicated to the destruction of human life, even though their actions may lead to precisely this result. By casting them in the dual role of promoter and regulator, we forced their actions to be evil even when their intentions may have been good. One only has to observe the reflex decerebration manifested by the Atomic Energy Commission or the Joint Committee on Atomic Energy when the words "death" or "cancer from radiation exposure" are mentioned to appreciate how inappropriate this dual role is.

Criticism of such decerebration is no more indicated than is criticism of reflex behavior of a dog conditioned to salivate at the ringing of a bell. And this leads us to consideration of more rational approaches to development of technologies like atomic energy.

The conflict between the Gallant Knight and his experts and the public

Our endeavor above has been to show that promotion of technology by a super-agency, such as the AEC, is an all-consuming affair, leaving little or no room for an ability to contemplate the consequences to human beings of the technology. Those who desire not to be victimized by the rashness of the entrepreneurial approach are required, in what has to be the acme of injustice, to prove that the technology has harmed or will harm them. The Atomic Energy Commission is not required to prove that their emission of radioactive poison is safe. Far from it! An individual who wishes to raise a question about safety of his exposure to radioactive poisons must use his own personal resources of funds for legal counsel and legal procedure. Arrayed against him is the entire Solicitor General's office of the U.S. Government and the vast resources of the Atomic Energy Commission to "purchase" testimony from hangers-on whose resources of research funds and livelihood derive directly from the AEC.

And if this lopsided array were not sufficient, the added insult is provided by the Judiciary Branch of government. The judiciary, as recent decisions indicate, operates on the presumption that the government would surely not set "tolerance" standards that are unsafe for the individual. Why and how this strange optimism on the part of the judiciary has arisen escapes understanding, but it is a fact of existence.

In summary then, an individual can be denied life, liberty,

and the pursuit of happiness by a government super-agency, and if he complains of the abrogation of his constitutional rights, he is likely to find major branches of government, with fantastic resources, arrayed against him with the purpose of denying him those rights.

Safety record claims are false

For years now we have been hearing of the inordinately good record of the atomic energy industry with respect to safety and freedom from fatal accidents. (Just because the AEC commissioners are making such assertions, we should not automatically assume that the assertions are false). However, even minor probing is sufficient to reveal that the assertion of a good safety record is not only false, it is absurd. It turns out that the record is "good" because it is so defined. Defining a record as good is achieved by the simple expedient of vigorously denying culpability for deaths that are obviously the direct result of the exposure to radiation of the industrial employee.

Let us consider the case of an atomic energy industry worker who received, over a 10-year period, the amount of radiation labelled as "tolerance." This he is legally allowed to receive. Our estimates, in good general accord with those of the respected International Commission on Radiological Protection, would indicate that after a latency period of 5-10 years or so, 1 out of every 2 cancers occurring in such workers are the direct result of the occupational exposure. So if we observe 100 cases of cancer or leukemia in such workers, the present evidence indicates that approximately 50 of them are occupational. By the remarkable expedient of defining a radiation dose which doubles the cancer rate as "tolerance," the atomic energy industry absolves itself of responsibility for these cancers. Thus, by simple definition, the atomic energy industry has an excellent safety record, even if it produces thousands of fatal cases of cancer and leukemia.

Technology sings its self-praises of the wondrous benefits it is conferring, or is about to confer, upon the unwitting population. When pressed, a technological super-agency, such as AEC, adds to its repertoire of lullabies about benefits a new tune, entitled "The Benefits Outweigh the Risks." As sung by the AEC Commissioners, the ditty is meaningless and, of course, they have no intention that it shall be otherwise. But within these words is the germ of an idea that can be the basis of a rational approach to technology and its associated poisonous by-products. Society may indeed require the benefits a new or existing technology has to offer. Further, society may find itself in a position where it is willing to accept certain grave risks in exchange for receipt of the benefits. It can be said, without any fear of contradiction, that society has never been given the opportunity to do so for three major reasons: (1) The benefits have always been vaguely described, at best; (2) the risks have been denied, minimized, or lied about; and (3) no weighing of benefits against risks by society has even been approached, since self-styled "expert" groups have made any such (dubious) calculations of this sort for society.

The flagrant disregard for the primacy of human health in such matters is beautifully illustrated in the 1967 hearings before the Joint Committee on Atomic Energy on the subject of the outrageously high lung cancer death rate being caused in uranium miners by exposure to radioactivity in the mines. We quote directly from The Federal Radiation Council Report No. 8:

Factors Related to the Evaluation of Benefit and Control Capabilities

Available information on benefits to be derived from the mining of uranium, difficulties encountered in reducing radon daughter concentrations from previous levels to current levels, and the additional difficulties that can be anticipated if further reduction in radon daughter concentrations are required has also been reviewed. The findings of immediate interest are as follows:

1. Uranium is currently the basic fuel needed for the development of nuclear energy, and all projections point to an increasingly important role for nuclear energy in meeting national electric power requirements.

2. Uranium mining is an important economic asset to the

States in which the ore is mined. In addition to the value of the ore, mining provides important opportunities for employment. It is estimated that the work force will vary between 2000 and 5000 men in the next decade.[11]

Stripped of euphemisms, the Federal Radiation Council staff appears to be saying it would be tragic to have nuclear fuel (uranium) cost a little more just to keep uranium miners from an epidemic rate of lung cancer. After all, the country needs electric power. Further, it helps the economy of the mining states, and if we make the mines safe, business might decline.

The final result in this magnanimous weighing of benefits versus risks occurred when the redoubtable Atomic Energy Commission recently awarded a $200,000 contract to the Arthur D. Little firm "to study the *economic impact* on the uranium mining industry" if they were forced to clean up the mines to a point where the lung cancer epidemic among miners might be mitigated somewhat. This is not a reasonable manner in which to manage the affairs of a great nation.

And all of this teaches us, and should teach the Congress, a most important lesson in the effort to preserve a livable environment for human beings with respect to radioactivity or other pollutants: Expecting scientists or other experts whose research funds and livelihood come from a promoter of technology to provide the truth concerning hazards, where the truth thwarts the technology, is like expecting our Christmas Eve dreams of sugar plum fairies to become reality. Sugar plum fairies may be real, but we better not count on it.

4 Protection policy against future pollution

Environmental pollution is a matter of extreme moment. Decisions concerning pollution should not be made in secret by so-called experts. The burden of proof should be shifted from the public and/or the government regulatory agency to the polluter. The polluter must be made responsible for convincing the public that he has done everything possible to reduce the level of pollution and that the benefits to be derived from his activity outweigh the risk of the remaining pollution.

Pollution and the fragile human organism

Mankind seems to have an unbelievable amount of self-esteem. We believe that we can take a tremendous amount of adversity and survive and in this belief, we are correct. But the important fact that we seem to overlook is that we pay for these insults to our physiological competence. We pay for them in terms of reduced physical fitness and a shortened lifespan.

For the wide variety of toxic materials that are introduced into our environment as pollutants, there are various standards established that are called permissible levels or maximum permissible levels. Generally, these standards represent concentrations below, usually considerably below, the level where immediate and obvious symptoms of disease would occur. We are, therefore, lulled into complacency by being led to believe that concentrations below this permissible level are harmless.

This is not necessarily true. In fact, for most pollutants it is undoubtedly incorrect. Although it is below its permissible level, a pollutant is most likely still causing its adverse effect, but at a rate that was too small to observe in the small number of short-lived experimental animals upon which it was tested, or in the brief period of time that it was tested in a small group of human subjects. The human subjects are usually adults and little is known about the long-term effects on the growing and developing child. As a result, the pollutant may have an effect that was overlooked in the testing procedures or could not have been observed in the tests. Such would seem to be the case with thalidomide and, as a result of that disaster, new drugs are now tested for their effect on the developing fetus.

Moreover, the effect of two pollutants in combination may be far worse than the sum of the effects of the individual pollutants. For example, radiation combined with cigarette smoking is ten times worse than radiation alone. It appears more likely that this synergism among pollutants will prove to be the rule rather than the exception. We should seriously consider such statements as those of Dr. Umberto Saffiotti, Associate Scientific Director for Carcinogenesis, National Institute of Health. "The striking potentiation of effects of low levels of a systemic carcinogen in the lung by as simple a treatment as the pulmonary penetration of a dust warns against the dismissal of any carcinogenic exposure—even at low levels—as being 'safe'."[12]

It must be remembered that even a food additive is a potential pollutant and could have a small adverse effect on every individual, or a serious adverse effect on 1 in 10,000 individuals. Either effect could have been unobserved or unobservable in the testing procedures. Either effect could cause a large amount of injury when, aided by mass distribution and mass communication advertising, the product is made available and attractive to 200 million individuals. Former Secretary of Health, Education, and Welfare Robert H. Finch's decision on cyclamates was a courageous departure from the past and an essential step into the present.

The point we are trying to make here is that the uncertainties connected with the effects of radioactive atoms are shared by practically every form of environmental pollutant. We are most likely paying a price for each pollutant, and the net effect of all of them may be more than we would like to pay.

Why do we have pollution?

When we survey the arsenal of scientific and technological knowledge that is available to this nation and its industry, it is obvious that the means are available to essentially eliminate all form of environmental pollution. There is one exception to this, and that is waste heat. We will return to this problem subsequently. There are numerous signs today which demonstrate that the present levels of pollution are detrimental to man and his environment.

The developing nuclear electricity industry in the country offers a current example of why we have such a serious pollution problem. At the same time, we can and should learn from this industry what is required to improve the quality of the environment and the quality of life in this country. This industry is at the heart of the problem because, in addition to being a polluter itself, it will generate the power to operate other industrial polluters.

As long as there is a legal limit or no limit to pollution, any nonsensical industry can pollute. A legal limit to pollution either implies that there is a safe level of contamination or that the process of polluting has a benefit to society that outweighs the attendant risk. We have no evidence whatsoever to indicate that there is a "safe" level for any form of pollution. Moreover, when a *legal* limit is established, pollution occurs without any balancing of benefit vs. risk.

The AEC suggests that it has done a risk vs. benefit calculation and has found that the benefit outweighs the risk. But, they never present a benefit value, and they detest people like us who dare to present a risk value. Consider the statement by Dr. Glenn C. Werth, Associate Director for Plowshare* at Lawrence Radia-

*Plowshare is a program of the U. S. Atomic Energy Commission. It is managed by the Division on Peaceful Uses of Nuclear Explosives

tion Laboratory commenting on a question posed by Senator Mike Gravel:

> It is difficult to balance a risk of radioactivity against a benefit. There is a need for natural gas. One of the most thorough studies is that by the Federal Power Commission entitled "A Staff Report on National Gas Supply and Demand," Bureau of Natural Gas, Federal Power Commission, Washington, D.C., September 1969. If more gas were available, it could be burned in more cities and significantly reduce the smog and health hazard associated with the presence of smog. Balancing the health hazard due to smog against a possible health hazard due to background levels of radioactivity has not been done to my knowledge.[13]

Why don't we do this study before spending millions of dollars on the gas stimulation program? Would such a study show that piping radioactive gas into homes is a reasonable solution to the smog problem? It would seem that even Congressman Holifield doubts the risk versus benefit calculation in this case because he asked why 50,000-million cubic feet of gas should be shipped to Japan each year if the shortage of natural gas were as serious as the AEC has indicated.

After you listen to their arguments for a second time, if you are not too terribly naive, you realize that all they have done is a cost analysis. For example, nuclear reactors are only marginally competitive with fossil fuel plants today. Any additional restrictions would price them out of business. When we say business, we mean big business. The bill for the present reactors exceeds $25 billion. The industry would like to increase that 2 or 3-fold. Because of the size of the market, some nuclear critics are accused of being in the employ of the coal industry.

All the nuclear critics that we know deplore inadequate fossil fuel generating plants as much as, and even more than, nuclear plants. No one can deny the ill effects of the noxious gases that

---

(D.P.N.E.). This program includes the study of peaceful uses of nuclear explosives, such as digging canals, excavating harbors, making mountain cuts, stimulating natural gas and oil production, and facilitating metal ore mining.

belch from the chimneys of these fossil fuel plants. And this is why society as a whole must become involved in this controversy. If the fossil fuel plants are forced to remove their noxious gases (which they can undoubtedly do) their cost will increase. The nuclear plants can then meet more restrictive controls and stay competitive. The question which has never been seriously considered is whether or not members of society are willing to accept an increase in their electricity bills. Strangely enough, all indications are that they would.

But noxious gases and radioactivity are not the only by-products of electric power production. There is waste heat. Enough waste heat to change our ecology drastically if our projected power needs are real. Consequently, public discussions must not be restricted to, for example, at what temperature shall the heated water from a given plant be discharged into the public waters or how much radioactive waste shall be discharged into our common air supply. To begin by asking these questions is to begin in the middle of the story. We must start with the fundamental question.

Do we really need more power?

What then is the fundamental question involved with the electric power industry? It is, *"Why more power?"* This question has not been publicly discussed until very recently. A flat and unqualified statement that ". . . power needs are doubling every eight years" is not sufficient. To accept this statement without question is to accept and endorse the notion that electrical power consumption is a desirable end in itself. Today, when environmental questions are paramount, it becomes necessary to question the basis for all intrusions on the environment. We do not know that we need more power. The population of the United States increases at about one percent per year. It is certainly not obvious that a population increase of one percent per year demands an increased power consumption of about ten percent a year.

It is manifestly not obvious that power *demands* are equivalent to power *needs*. How is the power to be used? Our utility friends advertise the use of power for lighting hospital operating rooms, running audiovisual aid equipment in elementary schools, making possible stereo recordings of Brahms and Beethoven and a host of other culturally interesting uses. It is highly unlikely that these uses account for a significant fraction of the present or projected power use. If we look closely, we will probably find that the Pacific Northwest needs more power to operate aluminum smelters in order to meet the growing "need" for beer cans and TV dinner trays. We must face the unfortunate fact that power consumption today does not correlate with the nebulous "standard of living." Power consumption is correlating with the production of garbage and the decline in the quality of the environment.

Americans have long recognized and taken pride in the beauty of their country and the nature of their democratic society. Beginning with the presidency of Theodore Roosevelt, we began to recognize that unless we changed our course of action, beauty would not persist. This was the beginning of conservation as a national policy in the United States. This represented a manifestation of the moral attitude of U.S. citizens concerning their desire to preserve this great land for future generations. Our efforts in the conservation field have always left a gap between action taken and action needed. This gap has grown so that we have now a great awareness of an impending environmental crisis.

Saving the earth for future generations

Conservation of our environment as a worthwhile national goal is supported by the overwhelming majority of our citizens today. This opinion still has its roots in the moral desire to preserve the world for future generations. Its major adversary has been the belief in the omnipotence of science and technology. Considering the rapidly deteriorating environment, one might hope this belief is losing ground.

By an accident of history, this growing awareness of the deterioration of our environment is occurring simultaneously with the extremely rapid growth of the nuclear industry—an industry that simultaneously offers the hope and the ultimate peril for future generations. The radiation emitted by the nuclear industry can influence the genetic makeup of many future generations. It asks the question: "Will your grandchild's genes be fit for your great grandchildren to wear?" This is really the ultimate in moral decisions.

Considering the morality of the great majority of the United States citizens, we doubt if they would accept anything other than the smallest possible change in their gene pool. We firmly believe that their morality would cause them to vote for a doubling of their electrical power cost rather than suffer a significant genetic change. At the very least, we think their morality should be given its chance at the ballot box. Furthermore, we feel that they should be fully informed on the facts and assumptions so that they can make this judgment. They are not, as many would imply, too stupid to do it! If they are, our democratic system is a fraud and this we simply do not believe.

A recommendation for pollution control

This then brings us to the means of controlling pollution. The reason we have pollution is that it is permitted either by law or by the absence of law. As was stated earlier, if there is a legal limit or no limit to pollution, any nonsensical industry can pollute. A legal limit to pollution either implies that there is a safe level of contamination or that the process of polluting has a benefit to society that outweighs the attendant risk. We have no evidence whatsoever to indicate that there is a "safe" level for any form of pollution. Moreover, when a legal limit is established, pollution occurs without any balancing of benefit vs. risk.

To properly protect the public health and safety, the laws should read that the acceptable limit of pollution is *zero* and that the privilege of releasing a pollutant to the environment must be negotiated. The prospective polluter should be required to demonstrate in a meaningful manner that his activity will produce benefits to those affected that outweigh the risk.

This weighing of benefit versus necessary risk should occur in public hearings before pollution control boards. It is important to emphasize the word *necessary*—the benefits must be weighed against the necessary risks. The right to overrule a decision of the control boards should be reserved for the public through the courts or by referendum. And this right must be reserved permanently so that any prior pollution allowed can be stopped if the public so decides.

Pollution tolerance should be zero

For questions such as the preservation of a livable environment and a livable world, it is truly discouraging that the public must plead for an opportunity to be represented in decision-making. No right is closer to the constitutional guarantees than that of a livable environment. Long ago the burden of proof should have been placed upon the perpetrators of pollution rather than upon his prospective victim. It has been the lack of the zero tolerance limit for pollutants that has led to the present serious impasse.

We have previously explained how useful the "acceptable tolerance" limit concept has been to—and so carefully cultivated by—the promoter of technology. The "acceptable tolerance" concept never had any scientific basis; it was purely the result of opportunism plus ignorance. It should not be necessary to fight for the abolition of this Neanderthal concept of treatment of the environment and its inhabitants. Zero tolerance has *always* made sense with respect to pollutants, and yet when it is now proposed, promoters and their disciples view it with disbelief. "We've always had a 'tolerance' limit," they argue. Indeed they have, and this is precisely why we have an environmental crisis. "Do you want to stop technological progress?" they ask. Our answer is, "No, we simply want technological progress to serve society's needs rather than to serve itself!"

Circumstances may very well arise where the hazard to society may be greater by impeding certain technologies than by allowing them to proceed. The concept of determining if benefits to be received by allowing the privilege of pollution does indeed outweigh the risk of injury from the pollution is a meritorious concept. The polluter, or potential polluter, speaks glibly of this benefit-risk calculation, but he has yet to perform the calculation with real numbers. He is quick to state that his particular pollution at the "tolerance" level has never yet been proved to produce harmful effects. Three questions then are particularly relevant: (1) Why is it essential that the activity he proposes requires *any* release of pollutants? (2) Why as much as the "tolerance" level? (3) What effort has been made to determine the effects, short and long term, on human beings and the ecology of pollution at the "tolerance" level?

Promoters convert 'demand' into 'need'

Almost invariably the answer to the first two questions is that it would be "uneconomical" to reduce the extent of the pollution. Therein lies the essence of the difficulty. What needs to be asked is "Economical for whom?" For the vast majority of instances, if not all, it is the economics of instant greed and gain versus the "economics" of quality of life for large numbers of humans. Since so very many of the products of industry are produced because they can be sold, rather than because of anyone's need, the most economical approach in the larger sense would be to prevent the particular activity altogether. The promoter can shuttle back and forth between the terms "demand" and "need" with great skill. His forte is to concoct a demand, refer to it for a while as a demand, and make the subtle transition to a desperate need.

And the second question concerning the effort to determine the effects of the pollution at "tolerance" levels is almost always answered as, "No effects have been observed in spite of careful observation." Careful observation of what? Generally, on deeper probing, it turns out to be a careful observation of nothing at all. If indeed any observations were made, they are usually irrelevant for the problem at hand. It is next to hopeless, thus far, to get science and technology in the service of polluters to understand the difference between no effect and "no effect observed." But it is absolutely imperative that this difference be stressed again and again by the public, for the experts will never do it. The public understands quite well that if an inadequate study is done, no effects may be observed. But the effects are there nonetheless and they can be devastating.

Even after the concept of zero tolerance for pollutants with negotiation of the privilege to deviate therefrom is codified into law, the public should expect that the risk assessment by the polluter and his experts will always be lower than the real risk. Part of this will result from inadequate search for effects; part will result from the gaps in our biological understanding at any point in time. And for these reasons risk estimates should always be suspected, should be reviewed with great care in an honest open forum, and should be subject to review in the light of new knowledge.

Today tolerance limits of pollution represent nothing other than a hunting license for human beings. Even with zero tolerance levels coupled with a specified deviation by negotiation, it is vital for the public to retain the privilege of revocation at any time that the benefit versus risk balance changes in the light of new evidence.

5 Lip service to the public health

"The Lawrence Radiation Laboratory will soon be announcing radioiodine is good for babies," we were told by I. F. Stone's *Bi-Weekly,* issue of June 24, 1963.

"The Atomic Energy Commissioners are on the hot seat, and something has to be done," we were also told by Dr. Spofford G. English, an assistant general manager of the Atomic Energy Commission in the spring of 1963. These words represented the basis for an exploration of the possibility of setting up at Lawrence Radiation Laboratory a comprehensive long-range program of investigation of effects of AEC programs upon man and his ecosystems.

Why were the AEC Commissioners on the "hot seat" in 1963, some *18 years* after the establishment of the AEC through the Atomic Energy Act of 1946? What special events signalled difficulties for the AEC? An answer requires a brief consideration of the prior history of atomic energy development.

The LRL Bio-medical Program and the Plowshare Program

Few controversies have been more bitter throughout the world than that which developed during the 1950s concerning the damaging effects upon humans of radioactive fallout from nuclear weapons testing by super-powers. Excellent studies by Edward B. Lewis, Jack Schubert, Alice Stewart, and Linus Pauling,* and those of a host of geneticists, provided abundant

---

*Dr. Edward B. Lewis is Professor of Biology, California Institute of Technology. Dr. Jack Schubert is co-author with Dr. Ralph E. Lapp of

reason for concern over possible irreversible and massive deleterious effects upon present and future humans. That the vast bulk of the world biological community shared the concern of such leading scientists is well known. Few would doubt that this concern was *the* major influence in ultimately leading to the Treaty to Ban Nuclear Tests in the Atmosphere. And that such concern led to the richly deserved award of a *second* Nobel Prize to the outstandingly great chemist, Linus Pauling, this time for contributions to peace, is appreciated throughout the world.

But, unfortunately, a rational solution to the problem created by the radioactivity and radiation associated with atomic energy has not yet been reached. Indeed, as shall be developed here, the problem has been made worse by the burgeoning of the "peaceful" atom, which now bids fair to compete successfully with the "warlike" atom in creation of human misery, death, and even possible human genocide.

The decade of the 1950s saw the extensive testing of nuclear weapons in the atmosphere, on the sea, and on land, primarily by the United States and the USSR. The cold war was in full bloom, and prevailing philosophy was that life could only continue as a balance of nuclear terror between superpowers. And the leading militarists of the superpowers, with scientists in collaboration, singlemindedly pursued the concept that survival was contingent upon staying one step (or more) ahead of potential enemies, in the quantity, deliverability, and sophistication of nuclear weapons. This concept led directly to massive testing programs of such nuclear weapons.

With the discovery that massive quantities of radioactive substances, by-products of nuclear explosions, were falling out over virtually the entire globe, being concentrated in foodstuffs, such as milk, and thus entering the bodies of humans and numerous other species around the world, public indignation,

Radiation: What It Is and How It Affects You. New York: Viking Press, Inc., 1958. Dr. Linus Pauling is Nobel Laureate in Chemistry and Nobel Laureate in Peace. Currently he is Professor of Chemistry, Stanford University.

fear and concern grew. Such concern represented a thwart to continuation of the military program and, therefore, in the military mind, required allaying of the public fear.

In the United States this task fell to the Atomic Energy Commission, in particular its Division of "Operational Safety." So far as can be ascertained, the only contribution of this division, or of the AEC itself, in this area was simply platitudinous reassurance. It is hard, in retrospect, to have expected much more than reassurance, since the distribution, by spewing, of radioactivity into the biosphere we now know can hardly have been expected to produce anything other than human misery and death.

The AEC's tranquilizer technique

Early in this reassurance effort the Atomic Energy Commission learned a temporarily useful technique in allaying public fear. Assume some amount of radioactivity is safe, even though no evidence exists for such safety, and repeat over and over that amounts of radioactivity released have not exceeded the "safe" limit. Perhaps the most amazing thing of all is that this supreme falsehood worked as a partial tranquilizer for as long as it did. As the concern of outstanding biologists grew, public fears grew, and no doubt these were major factors, among others, that finally led President Eisenhower to a unilateral moratorium on atmospheric nuclear testing in 1957. The USSR also ceased nuclear weapons testing in the atmosphere.

But in 1961 the Soviet Union resumed large scale nuclear weapons testing in the atmosphere. The United States quickly followed suit, and the radioactivity pollution of our atmosphere mounted. In 1962, the even more insulting phenomenon of atmospheric testing within the continental United States at the Nevada test site was again carried out, both for national defense purposes and for the now-infamous program known as Plowshare, dedicated to developing "peaceful" uses of nuclear explosives. By now, however, a modest sophistication concerning the long distance spread of radioactivity from such tests had developed. Measuring stations were available to test for fallout of a variety of radionuclides, including their appearance in such vital foods as milk.

Thyroid cancer from contaminated milk

The State of Utah has been mercilessly clobbered by nearly all such atmospheric tests in Nevada because prevailing winds generally carried the radioactive debris there. One radioactive substance released is especially pernicious—that known as radioiodine. This radionuclide, deposited on forage, is taken up by what amounts to a carpet-sweeping action by cows, and concentrated in the cow's milk. Infants and children drinking such radioiodine-contaminated milk extract the radioiodine from the milk and cencentrate this poison in the thyroid gland, a gland of very small size which has an affinity for iodine. The final result is intense irradiation of the thyroid gland by the radioactive decay of the radioiodine. As we have learned to our sorrow, radiation of the thyroid gland in children leads to one ultimate result, thyroid gland cancer—15 years later.

The Atomic Energy Commission, with characteristic aplomb, presumed that all the old clichés would work as they had before —platitudinous reassurance. Its Division of Operational Safety had ready for presentation its lullaby, "The Atomic Energy Commission conducted nuclear explosives tests in Nevada. No harmful quantities of radioactivity were detected offsite." One needs to understand the AEC's concept of "harmful" in order to appreciate this ludicrous statement so routinely dusted off and presented. When the AEC says radiation is "harmless," it means that people don't drop over dead by the droves as a result of exposure. As for cancer or leukemia developing 10 or 15 years later, who'll know? Who will ask? This is the "harmless" radioactive dosage as the AEC sees it.

But the situation in 1962 *was* different. A network of testing milk for radioiodine existed and was in operation in the State of Utah. And, as a result, the lies of the AEC concerning no harmful amounts of radioactivity being released in the Nevada tests were quickly exposed. Limits (which we now know are by no means safe) prescribed by the Federal Radiation Council were in danger of being exceeded in Utah milk, and the State of Utah was in an uproar and seething with highly justifiable indignation.

And the Atomic Energy Commissioners heard plenty about the radioactive iodine contamination of Utah milk. And that is why, in the spring of 1963, they felt themselves to be on the "hot seat." They simply had no good answers. In retrospect one wonders why anyone expected magic from the AEC Commissioners. Clearly, if atmospheric nuclear explosive tests spew radioactivity around the landscape, and if someone measures for such radioactivity, Houdini himself couldn't have made the radioactivity disappear. If no one knows enough to measure, which was the case in earlier years, it is possible to slip by unnoticed. But 1962 was too late for that system of deception.

The last thing anyone dreamed of was stopping atmospheric nuclear tests, and certainly the Atomic Energy Commission and the Department of Defense were thoroughly opposed to this solution. Something *else* must be done. In America we have a time-honored approach to nasty problems that refuse to remain submerged. We announce, with great solemnity and due publicity, that "We Shall Study This Grave Problem."

So, the AEC decided to approach the two nuclear explosive laboratories (Los Alamos* and Livermore) about doing something to get the Commissioners off the uncomfortable "hot seat." Their reasoning was straightforward. Nuclear explosives are designed and tested either by the Los Alamos laboratory or the Livermore laboratory. Nuclear explosive tests get AEC Commissioners in trouble with an indignant, fearful public. Perhaps if a group of biologists were working with nuclear explosive designers, either less radioactivity would be released from the explosions or the tests would be done differently. In any event, a good publicity release that the problem was receiving serious

---

*Los Alamos Scientific Laboratory is operated by the University of California under contract with the Atomic Energy Commission.

attention would certainly help *some*. The Los Alamos laboratory declined the offer to undertake the new responsibility.

Dr. John Foster, then director of Lawrence Radiation Laboratory at Livermore, was approached by the AEC with a request to consider setting up a bio-medical program in association with his nuclear explosives laboratory—for the purpose of studying the grave problem of fallout from such tests. He called us in to ask for an opinion concerning the wisdom and necessity of setting up such a bio-medical program. Was such a program necessary? Could it make a national contribution? Who would be available to do the work if it were to be done?

Thorny questions concerning radioactive fallout

Both of us agreed that radioactive fallout from weapons testing and from so-called "peaceful" nuclear explosives, as well as other atomic energy activities, was indeed of grave concern. There *were* many unanswered questions, such as who was getting exposed to how much radiation? And what precisely were the effects to be anticipated? And could anything be done to make all this safer? Some 19 laboratories were already studying biological effects of radiation and fallout and related subjects under AEC auspices. Why weren't they providing all the answers that were required? Why set up still another bio-medical laboratory to study the radioactivity-radiation hazard problem?

On careful thought we decided it was desirable to have one laboratory consider, in a broad, overall manner, the impact upon life and society of nuclear programs, from the nuclear source (such as the explosive itself) all the way through air, water, soil, food chains into man, and the ultimate effect upon man and other members of the biosphere. Indeed, we not only thought it was desirable, but mandatory, after we realized that no one was taking a broad view of the impact of Atomic Energy Commission activities that spewed radioactivity around various parts of the world's landscape. But there were several very thorny problems. The past record of the Atomic Energy Commission was anything but encouraging with respect to objectivity concerning the health hazards of its programs. Why should one expect honesty and objectivity to arise suddenly?

What if one did take this assignment seriously and try to find out the true impact on man of radioactivity release from various AEC programs? And what if the results indicated that the best thing to do would be to cancel those nuclear programs? Could the Atomic Energy Commission be counted upon to tolerate such thwarts to its promotional efforts, just because of health hazards? Would we be allowed to tell the truth about whatever hazards we discovered, or might the frequently-used security classification "Secret" be slapped on so that the unsuspecting public would be none the wiser? Besides, the credibility of the Atomic Energy Commission was so low in the eyes of the biological scientific community that it hardly seemed likely that anyone would seriously believe results reported from an AEC-supported laboratory, even a new one, there being so little reason to have ever believed AEC safety pronouncements in the past.

A new program to study effects of radioactivity

Dr. Foster was most reassuring on all of these concerns. *This time, he felt, the AEC was in real trouble and that they were prepared to turn over a new leaf with respect to public responsibility.* And this was a nationally important problem. Besides, if the work were done at the Lawrence Radiation Laboratory of the University of California, the prestige of the University would be behind the work, and would also protect the researchers from being demolished by the AEC, even if the AEC were to be inclined to be harsh about release of the truth concerning radiation hazards. One thing the biological researchers could count on, he said, was that the Lawrence Radiation Laboratory would back them to the hilt under any pressure, and this would insure that the problems involved would receive a full, honest, open airing, let the chips fall where they may.

Wistfully, we now realize our question should have been, "Will you love me in September as you do in May?" For some reason, we didn't ask this, being gullibly trusting, or incredibly naive.

So, finally, after much debate and soul-searching, we agreed to plan a program of study for the Lawrence Radiation Laboratory at Livermore under the awesome title: "A Program to Investigate the Effects of Release of Radiation and Radioactivity Upon the Biosphere (Particularly Upon Man)."

And perhaps foolishly, because the issue for humans seemed to be important, we both agreed to give up what we were doing to establish this program of study. The Atomic Energy Commission was most eager, agreeing to essentially every stipulation in the program outlined. A meeting was hastily arranged in Washington to finalize the establishment of the new biomedical program at Lawrence Radiation Laboratory at Livermore. Chairman Seaborg and Commissioner Lee Haworth attended on the part of the Commissioners. Numerous members of AEC officialdom also attended, all enthusiastic about the new program to evaluate the impact of various AEC programs upon man and others of his ecosystem. One of us (Gofman), still skeptical with the memories of past AEC actions concerning biological hazards, felt a last clarification was worthwhile.

He said, "We think this program is important, worthwhile, and essential if atomic energy development is to remain or be consistent with the safety and welfare of the Public." "What," he asked, "would be the attitude of the AEC if the developing studies indicated any particular program of AEC were inimical to the health interests of U.S. citizens?" We indicated we wouldn't consider initiation of the program unless we were absolutely assured that no censorship, no suppression of biological hazard reports, no interference with criticism of proposed AEC ventures would be forthcoming. Chairman Seaborg and Commissioner Haworth responded quickly and forthrightly: "All we want is for you to tell the truth about biological and medical hazards. You need have no fear that there would ever be any interference with release of the truth."

The agreements were quickly reached to establish the program. Indeed, it is doubtful that any new program sponsored by the AEC ever got approved so rapidly as this one—a clear reflection of the urgency felt by the Atomic Energy Commission to demonstrate responsiveness to the issue of public health and safety, especially after the grand blunders in Utah during 1962. And immediately the AEC issued the following promising news release to newspapers around the country:

U. S. ATOMIC ENERGY COMMISSION
SAN FRANCISCO OPERATIONS OFFICE
2111 BANCROFT WAY
BERKELEY 4, CALIFORNIA

SAN NO. 339
TELEPHONE: THornwall 1-5620
Extension 212

ADVANCE FOR RELEASE
AFTER 7:00 A.M. (PDT)
FRIDAY, MAY 31, 1963

BIO-MEDICAL STUDIES PLANNED FOR
AEC'S LIVERMORE LABORATORY

A comprehensive, long-range program to explore in greater breadth and depth and sources of man-made environmental radioactivity and the effects upon plants, animals and human beings was announced today by the Atomic Energy Commission.

The program will be built up over a period of time as scientists become available to fill the developing needs for specific talents. The Livermore studies will place special emphasis on early fall-out—that fallout which follows a nuclear detonation within hours or days. The studies will become an integral part of the Commission's Biomedical Research Program and will be closely coordinated with AEC-sponsored research being conducted at many other Laboratories throughout the nation. The broad studies of world-wide fallout and radiation effects on living systems going on elsewhere will be continued.

The Commission has recognized the need for a central group which would plan and conduct studies of environmental contamination due to release of radioactivity from nuclear detonations conducted for peaceful or military purposes. At the request of the Commission a program has been outlined and is being established at its Livermore, California, site operated by the University of California's Lawrence Radiation Laboratory. During fiscal year 1964 it is estimated that the operating costs will be about $2,000,000.

Dr. John S. Foster, Jr., Director of the Livermore Laboratory, has appointed Dr. John W. Gofman to head the new program and to be an Associate Director of the Laboratory. Dr. Gofman is a Professor of Medical Physics in the Donner Laboratory on the University of California's Berkeley Campus.

The presence at Livermore of scientists knowledgeable in the ways man-made radiation is generated and released, plus the concentration of pertinent facilities, make Livermore unusually suitable for assessing the implications of the broad range of possible conditions under which such releases might occur. Included will be studies of short-lived and long-lived fission products and neutron-induced activity, with particular emphasis on their distribution and effects in living materials. In particular, the Laboratory will investigate the entire chain of events leading to radiation exposure of human beings which might follow. One of the early studies contemplated will be concerned with the short-lived fission product Iodine-131.

Biological and medical research will now be closely integrated with existing physical sciences programs of the Laboratory, as, for example, the Plowshare Program in which nuclear explosives are being developed for huge earth-moving projects, mining operations, etc. The existing highly-developed computer facilities at Livermore will be a vital aid in a continuing collation of the storehouse of research information already available.

♯ ♯ ♯

(NOTE TO EDITORS AND CORRESPONDENTS: This information is being issued simultaneously by AEC Headquarters in Washington, D. C.)

One veteran observer of the government scene, a bit hardened by experience, reacted to this AEC news release almost immediately. I. F. Stone in his *Bi-Weekly* of June 24, 1963 wrote: "The Lawrence Radiation Laboratory and The Atomic Energy Commission have just announced establishment of a long-range program to study the effects of radioactivity on man. The Lawrence Radiation Laboratory is the stronghold of Dr. Edward Teller (the father of the hydrogen bomb) and so, shortly, we can expect to hear an announcement from the Lawrence Laboratory that 'Radioiodine is good for babies.' "

The cynicism reflected by such a remark was typical of opinions prevalent at that time concerning the credibility of the Atomic Energy Commission concerning health matters. We of the Lawrence lab's new Bio-Medical Division winced a little at this remark. We knew it meant our job was going to be harder than ever, since people were likely to be unbelieving of any statements made by an AEC-supported laboratory. But we knew the problems involved were of desperate importance to the health and welfare of the country and, indeed, for human beings in general, and we were determined to do a first-rate job to evaluate the true potential hazards realistically and honestly. However, we now know that Mr. Stone appreciated far better than we did then the difficulty we would have getting the truth to the public without repression.

As we saw the problem, there were four essential aspects:

(1) Learning about *all* the major AEC programs that could be expected to be major sources of release of radiation and radioactivity and whether constructive suggestions could be made for minimizing radioactivity release at the source. Among these programs were nuclear weapons testing, "peaceful" nuclear explosives, the so-called "Plowshare" program, nuclear reactors for power generation and space flights, and radio-isotopes for industry and medical use.

(2) Learning in a systematic manner how radioactive pollutants would distribute themselves from nuclear events in the air, the water, the soil, and finally in foodstuffs ultimately consumed by man. And the final dose of radiation of man everywhere in the world, and over a long period of time, had to be known for each such nuclear event if, indeed, we were ever to evaluate the ultimate health impact of radioactivity release.

(3) Learning the effects upon this generation of humans and upon future generations of the accumulation of radioactivity in the various tissues of the body. For this generation our concern was with the so-called somatic effects, prominent among which are leukemia and various other forms of cancer. For future generations our concerns were necessarily manifold, including spontaneous abortion of fetuses, late fetal deaths, neo-natal deaths, and a host of diseases man is heir to because of his

genetic constitution. Indeed, as we now know medically, genetic inheritance, or our so-called constitution, plays a role in virtually all the important diseases of man.

(4) Countermeasure research. At every step along the pathway from nuclear source to effect of radioactivity in man, a potential exists for possible reduction in danger by cutting down radioactivity release at the source, by intercepting it at the food chain level, by procedures of increasing the rapidity of excretion of such materials from man's body and, finally, by attempting to prevent the effect on health of that radiation received.

The wonderful promise of nuclear explosives

Clearly, the most potent approach to the entire problem is to cut off the release of dangerous radioactivities at the sources, be they nuclear reactors, nuclear explosives, or radio-isotope shipments. But atomic energy was still at that time hailed as a wonder-child; nuclear explosive weapons were to protect us from would-be aggressors by the mechanism of balance of terror, "peaceful" nuclear explosives were going to do innumerable and remarkable tasks for man, including digging harbors, making cuts through mountains, digging interoceanic canals, breaking up natural gas formations to stimulate natural gas, and a host of other unnecessary tasks; and nuclear reactors were being developed for distant space propulsion, as well as for electric power generation here on earth.

No one in 1963 was seriously even thinking that the most effective way to cut the hazard was *not* to do the program leading to the release of radioactivity; programs were regarded as sacrosanct. But we realized that the time might come when precisely this recommendation ought to be made—stop that particular AEC program as dangerous to, *and* unneeded by, society. But such a recommendation, if made, had to be backed up by sound scientific evidence. 1963 was not the time for that.

We set about diligently to do work in all the areas just described. The scientific data required appeared endless, and neither we at Lawrence laboratory nor the officials of the AEC headquarters expected that the experimental work required would all be done at Lawrence. Indeed, it was expected that we would pull together information from researches, past and on-going throughout world-wide laboratories, identify crucial missing information, and do or get done those experiments required to fill gaps in our knowledge. All this started with much enthusiasm and high sense of purpose in the spring of 1963.

A number of events followed in sequence which gave us real pause in our concern about the seriousness with which the Atomic Energy Commission regarded the important mission they had assigned to us. Shortly after the Livermore Bio-Medical Program had gotten under way, one of us (Gofman) was called back to Washington for an "important conference" at the Division of Biology and Medicine (AEC) in Washington—a conference on the subject of radioactive iodine. The subject was certainly of interest, and directly a significant part of the considerations which had begun to occupy our attentions in the new work.

But it turned out that the purpose of the conference could hardly be designated as "scientific." What had occurred was that an employee of AEC Headquarter's Division of Biology and Medicine, a Dr. Harold A. Knapp,* had assembled available evidence from past nuclear weapons tests, and had concluded that the radiation damage received by children in the State of Utah from such tests had been *far in excess* of anything predictable from the past safety announcements of the AEC.[14] He wanted to publish his findings openly. AEC was extremely worried about the impact of his publishing these data. In essence, the message to our Committee assembled as a "conference" was, "How can we stop this report—a report which will, in effect, make the AEC reports over the past 10 years look untrue?"

Dr. Knapp was justifiably furious at the AEC efforts to block his report. To the credit of the Committee members convened

*Dr. Harold A. Knapp, Jr., is a member of the staff of Weapons Systems Evaluating Group, Office of Director Defense Research and Engineering.

to review the Knapp report, they recommended early publication of the Knapp report, in spite of AEC Headquarters' objections. Dr. Knapp published his report and resigned from the AEC Staff. We had our first direct taste of what Dr. M. King Hubbert* so aptly describes as AEC "sanitizing and cosmetizing" of scientific reports before release to the public. The scientific findings of Dr. Knapp that much more injury had occurred from radioiodine than previously admitted by AEC have never been refuted. Indeed, a subsequent report by Tamplin[15] confirmed and extended the findings, in essential agreement with Dr. Knapp.

Not long thereafter we came to learn more about the "seriousness" of the AEC desire to have a real evaluation of the true impact of radioactivity release upon man and the biosphere. Word filtered into Livermore that the Bio-Medical Program was in trouble. Why? Apparently in their haste to get something done to appease public indignation over the Utah fiasco, the AEC Commissioners had approved the program at Livermore in the absence of one member of the Commission, James T. Ramey.

A cut in the budget

Upon his return to Washington he was apparently furious that the program had been initiated without his approval. It was made clear to us that Commissioner Ramey disapproved of the program strongly and that there was the danger the program might be cancelled. Apparently it would have been too embarrassing, with the extensive AEC publicity about the program, to cancel it outright. Instead, the budget was cut even from the low starting value and we were given to understand, in no uncertain terms, that the program was *not* going to be supported at the level required to do the tasks outlined originally.

We were faced with a dilemma. The work, we felt, was of supreme importance to humans, to the public health and welfare. Several of us had uprooted our lives, our previous profes-

*Dr. M. King Hubbert is Research Geophysicist, U.S. Geological Survey.

sional associations, to undertake an important task in the national interest. Would we be able to continue the work? Was it worth trying to continue in the face of obvious lukewarm support in Washington—if not outright antagonism? Dr. Foster, then director of Lawrence Radiation Laboratory, was reassuring. We would have the support of the Laboratory and the University. The AEC wouldn't dare offend the University by simply cancelling the program outright, even with Mr. Ramey's outspoken antagonism. Besides, we were assured that after a while Mr. Ramey would get over his "pique" and become friendly to the program. But that was not to be.

The fight for a new bio-medical complex

The Bio-Medical Program at Livermore was at that time housed in makeshift quarters. It was anticipated and promised by AEC in the original discussions that a laboratory building would be budgeted and constructed with funds to be requested of Congress in the following fiscal year. The Joint Committee on Atomic Energy struck from the AEC budget the funds that were to be used to construct a Bio-Medical complex at Livermore. This was tantamount to the Joint Committee on Atomic Energy stating that the Livermore Bio-Medical Program was unnecessary, for without facilities to work in it was hard to envision much of a program being possible. Again it appeared that the highly vaunted essential program to evaluate the hazard of AEC releases of radioactivity—so "necessary" when the Commissioners were "on the hot seat" was no longer essential—especially since in June 1963 the atmospheric test ban treaty had been signed. No doubt the Commissioners felt a great deal of heat was removed by the elimination of that particular source of radioactive fallout.

Dr. Foster was not about to accept this slap at the Livermore laboratory, at least not without a fight. So, Foster and Gofman went to Washington to a hearing on the AEC appropriation before the Joint Committee on Atomic Energy. Chairman Seaborg testified that four of the Commissioners favored the construction of the Livermore Bio-Medical Laboratory building, but that Commissioner Ramey felt it was unnecessary. Mr. Ramey had clearly *not* given up his antagonism. Drs. Foster and Gofman testified about the importance of the bio-medical work at Livermore, its implications, and they urged strongly that the Joint Committee on Atomic Energy re-insert the budget item for facilities so that the work could proceed.

Why, precisely, the Joint Committee reversed itself and voted the funds for the Livermore Bio-Medical complex will probably never be known. Our opinion is that the Joint Committee on Atomic Energy did not wish to offend Dr. Foster, widely regarded as one of the top leaders in nuclear weapons technology. In any event, the Livermore laboratory Bio-Medical Program did get a building, it did get a begrudging lease on life, and was enabled to proceed with its work but under a severe budget limitation.

Reasons for budget limitations

The severe budget limitations were the result of two phenomena, neither of which had anything to do with the availability of funds to the Atomic Energy Commission. First of all, the "hot seat" of the AEC Commissioners had cooled off considerably with the signing of the treaty to ban nuclear explosions in the atmosphere. Radioactive fallout controversy would die down, or so it must have been hoped. Apparently the Atomic Energy Commission was having difficulty realizing that its several "peaceful atom" projects might ultimately deliver twenty times as much radiation as did the fallout from nuclear weapons tests, and that the cool seat might one day become sizzling.

The second reason for budget limitations was Commissioner Ramey. He had by no means given up his antagonism concerning the Livermore Bio-Medical Program's having been established in his absence. Every time the Livermore Lawrence laboratory directors complained to AEC officials about the failure to meet the prior assurances of budgetary support, they were duly informed that Mr. Ramey was still unhappy, and that it might be wiser to go ahead with the greatly reduced bio-medical program than to create a major scene which might ultimately result in no budget at all.

We accepted the reduced program, knowing that we could certainly still do much of the important work on radiation hazards.

The situation up to the present

As the work developed at Livermore, it became increasingly clear that the promoters of atomic energy technology were dedicated to one goal—bigger and bigger atomic energy programs, with truly little or no concern for the public health and safety aspects of such programs. The evidence which developed steadily and certainly lent credence to a recent statement by Dr. Peter Metzger* that "It appears that the Atomic Energy Commission has decided to cope with the danger from radioactivity by attacking the public fear of it."

In 1964 the Plowshare Program, dedicated to development and utilization of nuclear explosives for "peaceful" purposes held a symposium at the University of California at Davis. Gofman was asked to present a paper there on "The Hazards to Man From Radioactivity." By then the Bio-Medical Program had been underway a year. That paper was presented and it stated that we knew *far too little* about the hazards of radioactivity to comment on the loss in lives in this generation or future generations from the spewing of radioactivity that would necessarily accompany such projects as digging an interoceanic canal (a sea-level type Panama Canal) by the use of nuclear explosives. And further that the favorite cliché of atomic energy promoters, namely, that the "Benefits to be achieved outweigh the risks" was meaningless, since no evidence had been adduced concerning the benefits, and tremendous gaps existed in our knowledge of the true risks.

Plowshare advocates at Lawrence laboratory itself were

*Dr. Peter Metzger, biochemist, Ball Brothers Research Corporation; chairman of the Colorado Committee for Environmental Information.

furious about that presentation. Lawrence lab was a leading advocate of Plowshare, largely sponsored by Dr. Edward Teller, and almost the only technical center actively engaged in it. And here was a member of the same laboratory's Bio-Medical Program, and an associate director of the laboratory at that, saying our knowledge was too fragmentary to provide a bio-medical endorsement of the safety of that program. "Et tu, Brute?"

Indeed, the next day one of the leading exponents of Plowshare reported to Dr. Foster, the laboratory director, that the Lawrence Radiation Laboratory had a "Trojan Horse" in its midst. He told Dr. Gofman about this laughingly. It is clear, in retrospect, that failure to whitewash nuclear programs with respect to health and safety is certainly no way to win friends and influence people within the AEC "family." But that presentation of the "Hazards to Man From Radioactivity" did accomplish one major goal; namely, proving that an honest evaluation was not *totally* impossible from an AEC-supported laboratory.

Indeed, the highly respected Committee for Environmental Information in St. Louis, Missouri, requested permission to reprint the "Hazards to Man From Radioactivity" in its journal, entitled "Scientist and Citizen." As everyone actively following this field knows, the St. Louis group has performed a superb service over many years in bringing to public attention the truth concerning radioactivity, radioactive fallout, and their hazards to living creatures. For "Scientist and Citizen" to have carried that presentation is indeed a distinction, indicating credibility of responsible, concerned scientists.

'Peaceful' atomic explosions will spew radioactive poisons

While we didn't know the precise magnitude of the loss in lives and human suffering to be anticipated from a Plowshare Program, such as digging a Panama canal with more than 100-megatons of nuclear bombs, we certainly did know that the magnitude could theoretically range from a low number to some enormous cost in human lives. Such nuclear explosions, for "peaceful" purposes, represent uncontrolled spewing and dissemination of a variety of radioactive poisons in great quantities over lands and waters, there to be accumulated by plants, animals and man.

Reflecting upon this now, we realize we should not only have said that the hazards are unknown, but we should have also said that it represents utter folly to consider such projects precisely because the hazard to life could potentially be extremely large. In fact, Duane Sewell,* an associate director of Lawrence lab, asked Gofman at one of the directors' meetings what he thought biologically about constructing a new Panama canal by nuclear explosives by the year 1975. The reply was, "Biological insanity." That's more true now than ever.

Meanwhile Tamplin had organized a group to carry through one of the major tasks the new Bio-Medical Program had set for itself; namely, predicting over space and time, the dosage of radioactivity and radiation to be expected from any particular nuclear project, explosive or other, for people at various distances from the site of the project. Experience from the nuclear weapons testing of the 1950s and early 1960s had provided abundant evidence that winds can carry radioactive debris from nuclear explosions thousands of miles from the site of explosion.

Hence, it is evident that a responsible evaluation of the dosage to be expected for humans must be global in scope, and must address itself to consideration of every conceivable radioactive substance that could be released from, say, a nuclear detonation. Since this involves some 700 possible radioactive substances of varying lifetimes of existence and having widely differing properties concerning meteorological distribution and concentration in various members of man's food web, the enormity of the task is readily appreciated. But steady, effective progress in this large endeavor was being made by Tamplin and his Information Integration Group. Again, the atomic energy promotional philosophy raised its ugly head in criticizing the Tamplin approach. We must understand this problem.

*Duane C. Sewell is Associate Director (for Support) of the Lawrence Radiation Laboratory.

The Tamplin approach was to consider the health and safety of the public as paramount, and the promotional aspects of the nuclear project as irrelevant by comparison. After all, there is never a dearth of hoopla and promotional ballyhoo emanating endlessly from technological promoters, such as the U.S. Atomic Energy Commission. And to consider the health and safety of the public as a prime obligation it is essential to be conservative in estimates of what *might* happen in the project. For what *might* happen under adverse meteorologic conditions, such as a rainstorm intersecting a radioactive cloud, could lead to massively higher doses of radiation and radioactivity accumulation by people than the promoter's estimate, assuming everything would go off perfectly in accord with their hopes. The Information Integration Group expressed its philosophy well in the Preface to the many scientific reports on this subject in the following:

There are three questions that could be asked about the outcome of the detonation of a nuclear device. It is essential that the reader recognize which of the three we are trying to answer and why we feel it to be the most appropriate. The three possible questions are:
1. What is the worst situation that could develop?
2. What is the most likely situation that will develop?
3. What would be the situation if everything went off perfectly?

We choose to answer the first question and to direct our efforts to predicting the worse case. However, in the process of answering the first question we can generally answer the second. Quite obviously, the answer to the third question has no meaning with respect to public health and safety.

We choose to answer the first question because we feel that only the worst case should be compared with prescribed tolerances in a preshot rad-safe* analysis. Furthermore, only when we know the worst case can we establish an adequate system of postshot monitoring to document the actual case and to insure that appropriate countermeasures are instituted when and if needed. In other words, it is only by this approach that uncertainties concerning dosimetry, such as presently exist for I^{131}, can be eliminated.

---

*Radiation-safety

Furthermore, we are attempting a thorough analysis and are considering each and every radionuclide on the chart of the nuclides. In this respect, our estimates may indicate that a particular radionuclide is a hazard for one of two reasons: Either (1) it will be a hazard because of what we know about it, or (2) it will be a hazard because of what we don't know about it. If a pertinent relationship is not known for a particular radionuclide, we make worst-case estimates of the relationship and hence maximize our estimates of hazard. Nevertheless, despite its conservative nature, this approach still allows us to eliminate most of the raidonuclides from consideration and to indicate those that are potentially the most hazardous. Obviously, it also allows us to estimate the upper limit of the potential burden and dosage; but, due to the perversity of nature, the precise dosage can only be determined by postshot documentation in the affected areas.

It is obvious that had I^{131} been measured in milk during the early period of testing, its dosimetry would not now be a problem. Thus, we wish to be able through our predictive approach to indicate what should be measured, where it should be measured, and with what precision it should be measured. There appears to be no other way to insure unambiguous dosimetry for future events and to assure that the need for countermeasures is recognized in time so that they can be planned for and instituted when and if needed."[16]

(The reader might be interested in knowing that in 1970, after the current leadership of the Bio-Medical Division, Drs. Bernard Shore and Roger Batzel* began the reprisal destruction of Tamplin's research efforts. One of the earliest scientific casualties was the Preface recorded here, which explains responsible public health procedure. At long last, through the help of Dr. Shore and Dr. Batzel, the Preface, so troublesome to the Plowshare enthusiasts, could be hidden from further public view.)

Protection of the public from senseless, irresponsible irradiation was regarded as an absolute *must* by Tamplin and his colleagues. Therefore, recognition of the possibility that radioactivity levels might be higher in some locales than expected,

---

*Dr. Bernard W. Shore is Division Leader, Bio-Medical Division, and Dr. Roger E. Batzel is Associate Director, Chemistry and Biomedicine, Lawrence Radiation Laboratory.

and the readiness to impound contaminated food supplies in such an event, are the minimal elements of public health responsibility. A reader who asks why do such projects *at all* is, of course, thinking far ahead of the opportunism that characterized such technologies as atomic energy then, and now, for that matter.

It's the unusual cases that are revealing

The Tamplin "worst-case" estimates of radiation dosage that might occur were not drawn from "thin air." They were made utilizing actual experience from prior events where radioactivity *had* been spread around by nuclear tests. To be sure, such experience had not always led to major untoward results, but on numerous occasions it had. For public health purposes, it is such unexpected unusual cases that really need to be considered. If everything goes as planners hope, there never would be problems, since the planners don't plan for trouble.

Obviously, however, to protect the health of the public and to be prepared for preventive or remedial action, Tamplin's group necessarily had to assume the worst case. What a mockery it would have been not to anticipate the probability of higher-than-expected radioactivity deposition from, say, a Plowshare experimental nuclear explosion. A rainstorm can precipitate ten or more times higher-than-expected amounts of radioactivity out of a radioactive cloud drifting away from a nuclear detonation site. What protection would citizens have if such rainstorms were not anticipated?

But to the enthusiasts for Plowshare "peaceful" nuclear explosives, any suggestions that people might get exposed to excessive amounts of radiation represented a thwart of their "technological objectives." Success for the peaceful nuclear explosives program meant being able to explode thousands and thousands of such nuclear bombs, to dig many harbors, canals, to move many mountains. If the public became concerned about the prospect of being plastered with radioactive debris, there might, reasoned the Plowshare advocates, be a public outcry, a vigorous complaint to congressional representatives and senators. This obviously might cause restrictions upon the program, and such restrictions were to be avoided at all costs, no matter what the consequences to the health of the public.

So, because the Tamplin worst-case estimates concerned themselves with health protection of the public, his work became steadily more regarded by nuclear bomb advocates as anathema. And the bio-medical research at Livermore Lawrence laboratory came to be regarded by Plowshare advocates as the work of "the enemy within." When challenged about his "worst-case" estimates, Tamplin would point to case after case of nuclear explosions in the infamous Nevada tests where radioactivity deposition had greatly exceeded expectation—the expectation, of course, having been prepared by nuclear bomb advocates. The bomb advocates always had a ready explanation, "Oh, *that* particular test was 'atypical'." One has to understand this amazing use of an English word. For the Plowshare advocates, two phenomena make a nuclear explosion "atypical": (1) Any nuclear explosion where lack of foresight and planning leads to sufficient radiation exposure as to arouse public indignation, and (2) Any nuclear explosion where the public becomes *aware* that they may be clobbered by radioactivity.

Plowshare's 'atypical' results

There are some further insights into Plowshare enthusiasts' views concerning Tamplin's "worst case" estimates. In response to their constant chorus that, "This particular nuclear explosive test was atypical," Tamplin replied: "Of three recent Plowshare excavation experiments, you succeeded in making a crater one time, raising a hill the second time, and creating a volcano the third time. Which one was typical?" Needless to say, silence was the Plowshare answer.

Clearly, Tamplin's estimates, when made public, increased the chances that citizens might know what could happen and this was a thwart—hence, the appellation, "the enemy within." It was a depressing, but still interesting, phenomenon to observe as the bio-medical research at Livermore unfolded. Tamplin's group reports were regularly funnelled to the director's office at the Lawrence laboratory and a new round of haggling ensued every time over the Plowshare concern that the reports might lead to public concern or indignation. The work involved in getting a report released got to be nearly as extensive as the work involved in the scientific research and report writing itself.

Slowing down unfavorable reports

Finally, when "sanitization" of the reports was not easy enough for the laboratory Plowshare advocates, a remarkable expedient was devised. All reports pertaining to Plowshare must first go to Washington Plowshare—AEC headquarters for "review" and release. This clever procedure at least delayed possibly "unfavorable" reports concerning potential hazard of "peaceful" nuclear explosives. As an harrasment of responsible biological concern for the public, this procedure represented one of the most flagrant violations of AEC responsibility for public health and welfare imaginable. As a mockery of AEC concern for public health, a concern ostensibly manifested by its sponsorship of "health and safety" research, it was cynical in the extreme.

Wherever obstruction of such reporting of bio-medical concern did not suffice, a separate technique was available; namely, ridicule of the biologist's concern for the public health.

The old standard clichés were dragged out repeatedly. "Everything we have is due to technological progress," or "Do you want to return to the Dark Ages?" Hardly. We were *not* interested in returning to the Dark Ages. We didn't regard protecting the public health as a return to Dark Ages. We certainly favor technological progress, but progress that serves societal needs; not "progress" represented by unnecessary senseless spewing of radioactivity around the earth for all time.

One favorite project of the Plowshare advocates has been the excavation of a new Panama canal by the use of multi-megaton quantities of nuclear explosives. The bio-medical program was charged with the responsibility of considering the potential hazards of such nuclear excavation. Our earliest evaluations of the prospect of such a nuclear excavation were indeed grim, for the amounts of radioactivity to be released in an uncontrolled manner were astronomical. In retrospect, it is difficult to imagine any circumstances under which such excavations could make sense, but we were not quite sophisticated enough to realize this at the time. Instead we sincerely addressed ourselves to serious calculations of expected radioactivity releases and radiation exposure to humans to be anticipated if such a nuclear explosive digging of a Panama Canal were ever to be carried through.

In numerous ways we were informed of the cynicism and contempt held by some Plowshare nuclear explosive advocates for the bio-medical concern over the radioactivity associated with major proposed nuclear excavation projects such as the new Panama canal.

On digging a new Panama Canal with nuclear bombs

President Johnson had appointed a commission to study the proposal of a new sea-level canal, including such issues as whether it should be accomplished by conventional excavation techniques or by the blasting of nuclear bombs. In the sourse of that evaluation, General Kenneth Fields, a member of the Atlantic-Pacific Interoceanic Canal Study Commission, visited the Lawrence Radiation Laboratory to learn of work there concerning many aspects of the possibilities of nuclear excavation, including concerns over seismic (earthquake) effects and biological hazard problems associated with such use of hydrogen bombs. At a briefing session arranged for General Fields, Tamplin and Gofman were to present their studies and findings concerning the radioactivity hazards anticipated. Professor Edward Teller, a fervent and leading advocate of "peaceful" nuclear bombs, was also present, and he decided (without invitation) to provide a preamble to the Gofman-Tamplin presentation.

Professor Teller started by reassuring General Fields that there was no significant radioactivity hazard to be anticipated from the nuclear excavation of an inter-oceanic canal. Where this revelation came from we have no idea since we felt the problem required extensive and careful consideration.

Our speechlessness was further increased by Professor Teller's following views: In essence, he said it was all right for John Gofman and his colleagues in biology to do research to amuse themselves concerning biological hazards of radioactivity associated with a nuclear explosive digging of the canal, but there just was no hazard problem. Gofman, Professor Teller said, reminded him of the early days of the introduction of the railroad locomotive. Those who were worried, Professor Teller said, wanted a man on horseback preceding every locomotive, waving a red flag of warning. For the nuclear age, Professor Teller indicated that Gofman represented that man on horseback waving a red flag.

All we can say is that we, with a background in this field, felt there was much we had to learn. With no background of qualifications concerning bio-medical problems, Professor Teller's dismissal of the radioactivity hazard was naive at a minimum, and certainly a manifestation of the arrogance of a promotional technologist.

There is a fundamental error that characterizes the technologist in atomic energy, as well as in other technological programs. The Livermore technologists were no exception. With a powerful new technology in their hands such as that relating to nuclear weapons, the assumption is made that what they want to do with their technology must be good; therefore, the "benefits" certainly outweigh the risks inherent in the technology.

Technologists hope that risks will just go away

We have stated on many occasions that a valid, honest calculation of benefits to be anticipated versus risks to be expected (in premature deaths, genetic deformities, or other injuries) is highly desirable—mandatory, in fact, if the technology is being considered for use in any public widespread manner. The atomic technologists have by-passed this fundamental step. They *assume* the benefits to be superb and they *hope* the risks won't be severe enough to impede the wide application of their technology. It is a short step from this "hope" to an assumption that the risk is negligible. From there an even shorter step to antagonism toward any criticism of those who properly study and raise serious questions about risk of technology, atomic or otherwise.

What happens is that, known or unknown to technologists, they are so enamored of their technology that they make a value judgment that the technology offers society so much it just obviously is worth the risk. No more serious error could be made and none could cost society more dearly now and in the future. The price could be as high as the loss of the human species, as well as other members of the animal and plant kingdom.

Technologists faced biological unknowns

We felt the antagonism of various atomic energy promoters rise, both in our own laboratory and in Washington AEC headquarters with the progression of our work. Our mission was to raise questions concerning hazards and to endeavor to answer them. We knew the technologists simply had no reason to say they should be allowed to go ahead because we knew there were many biological unknowns. Statements from *advocates* of nuclear technology reveal that their knowledge of benefits versus risks is primitive.

We have previously (Chapter 4) alluded to the statement that Dr. Werth, Lawrence lab director for Plowshare, made concerning the difficulty of balancing the risk of radioactivity against a benefit. He was discussing the benefit-risk aspects of the Plowshare Program to stimulate natural gas with nuclear explosives. Unfortunately, the gas recovered is radioactive. Dr. Werth's statements were completely honest, and they demonstrate precisely why one should have grave reservations about this technology. For the director of the technology is saying there that he doesn't know that the technology is worth the biological price. Some of his associates are much less careful.

For example, in a paper presented by Dr. J. J. Cohen of the University of Michigan, at a Health Physics Society meeting, the following astounding statement is made: "One rem of radiation and its somatic and genetic effects are worth approximately $250."[17]

Elsewhere we have presented the number of extra cases of cancer and leukemia (somatic effects) and the hereditary diseases (genetic effects) to be expected from one rad (or rem) of radiation delivered to human beings. Let us consider just the somatic effects, cancer plus leukemia, and neglect, momentarily, the probably far larger genetic effects. We showed that the best estimate for one rem of radiation is that there will be a 2% increase in the occurrence rate of cancers + leukemia. Since there are about 300 cases of cancer plus leukemia for every 100,000 adult persons per year in the United States currently, this means that one rem of radiation will produce 300 x 0.02 = 6 *extra* cancer cases per year. In 20 years, this represents about 100 extra cases of cancer + leukemia. Since we are dealing with 100,000 people, we can use Dr. Cohen's figure of $250 per person, and arrive at $25,000,000 as his acceptable value for producing 100 extra cancer cases; or $250,000 per case of cancer or leukemia.

Setting a price on human life

This is the first time we have seen such a sanguine evaluation of the worth of a human life: the dollar value of the misery and suffering of a case of cancer or leukemia fixed at $250,000. Maybe our estimates are wrong; maybe it would come out $100,000 or $500,000. To whom is it worth $250,000 to have one person suffer this fate unnecessarily? To the victim? To the family he leaves behind? Have either the prospective victims or their families been consulted as to how they feel about donating a life of one of them so society can have $250,000? Is *this* human? Is *this* our concept of human morality? But this calculation emanates from an atomic technologist in a serious discussion.

We might point out to Dr. Cohen that since we only considered cancer and leukemia here, he is getting the possibly six times greater number of genetic deaths in addition to the cancer death for the $250,000. This "bonus" makes human life, or loss of it, even cheaper!

Other experiences we had with atomic technologists are even more illuminating. Professor Teller is a leading figure in the Lawrence Radiation Laboratory. Not too long after the establishment of the Bio-Medical Program at Livermore, Dr. Teller called several of us into his office. The subject of the discussion was what he thought a primary objective of the Bio-Medical Program should be at Livermore. We thought we had quite well in mind what our objectives were and what work was required. But certainly we were most eager to receive suggestions from everyone as to how to do our tasks more effectively—especially from someone with the nuclear technology experience of Professor Teller.

"President Kennedy," Dr. Teller related to us, "has stated in his 1961 inaugural address that nuclear war could result in the abolition of human life on earth." We thought this was a very important statement that President Kennedy had made, and that it should be made over and over again. And the more prominent people who make this statement heard worldwide, the better.

But Professor Teller was most disturbed by President Kennedy's inaugural address. He (Dr. Teller) had given very serious thought to the problem of how many people might survive a worldwide nuclear war; he was certain that *some* human beings would survive. The Bio-Medical Program, reasoned Dr. Teller, should set as a high priority the development of evidence that *some* human beings would definitely survive a nuclear war. And if, he said, we accomplished such a worthwhile endeavor, then perhaps on a second inaugural address President Kennedy would not again repeat the serious error of saying that no human beings would survive a nuclear war.

We left that conference numbed, speechless. What if 2 people,

200 people, 200,000 people, or 2,000,000 people survived a nuclear war? Should President Kennedy really have altered his statement? Needless to say, the Bio-Medical Program at Livermore did not incorporate these supremely important studies in its work. But we had learned a great deal more about the thinking processes and the considerations important to some nuclear "technologists."

Dr. Teller is unworried over nuclear weapons tests

Professor Teller's view that radiation is not a serious hazard at low or moderate doses is well-known to the world. He was expounding this view in his effort to convince the world that nuclear weapons tests in the atmosphere were not creating untold misery and death from nuclear fallout radiation. The reader may recall the famous debates on this subject between Dr. Linus Pauling and Dr. Teller during the 1950s. The Nobel Committee awarded Linus Pauling a *second* Nobel Prize, this time for Peace, as a result of his work on this subject. Dr. Teller was very unhappy that we abolished weapons tests in the atmosphere. And he is certainly still unconvinced that radiation is very worrisome.

When Professor Teller states an opinion, we should scrutinize it carefully—as, for instance, the following quotation from a letter to Senator Mike Gravel:

> The present guidelines for "Permissible Doses" should not be lowered for the following reasons:
> (1) On the basis of common sense the present guidelines are safe.
> The main reason for this statement is that the guideline coincides with the average exposure due to causes other than atomic energy developments. This exposure has existed for a long period and furnishes a strong link with experience.
> It is generally recognized that the danger to an individual is small if 0.17 rem/year is added to the existing average of 0.17 rem/year. The fact that the chance of damage is so small makes it most difficult to find and prove damage at these low levels of irradiation.[18]

These statements of Professor Teller must be regarded as a classic of our times. Let us explore his statement that "the danger to the individual is small" if he received 0.17 rem/year additional radiation. If everyone in the U.S. received this dose, there would in time be 32,000 extra cancer and leukemia deaths per year. For our country this represents an unparalleled public health disaster. Of course, since there are 200,000,000 people in the country, this means one person out of every 6,000 people per year as an additional potential radiation cancer victim. If one chooses to look at a major public health disaster as "small danger to the individual" simply because only one extra person in 6,000 dies per year, then Dr. Teller is *technically* correct. But this simply means he doesn't appreciate that major diseases kill 1 in 1,000 or 1 in 10,000 per year, and medical research and treatment go to great efforts to prevent occurrences of this frequency.

Radiation is harmful whether natural or man-made

Dr. Teller's statement that damage is "small" at existing levels of 0.17 rem/year is totally without foundation. Our evidence indicates that natural background radiation plus medical exposures which account for the existing 0.17 rem/year do *considerable* harm both in the production of cancer and leukemia and in production of genetic disorders. Just because we can't do anything to stop natural radiation doesn't mean it does no harm. There is absolutely nothing, nothing at all, from anywhere, that indicates natural radiation to be less harmful than man-made radiation. A gamma ray is a gamma ray, no matter *where* it comes from. Professor Teller knows this very well. Why he or anyone would separate natural radiation from other radiation is beyond us.

And how he assumes, through common sense, that man polluting himself with an additional amount of radioactivity and radiation equal to what he already gets is *safe* defies comprehension for us, and probably for most of the world's biological community. These biologists worried seriously and properly

Lip service to the public health 83

along with Linus Pauling about adding fallout radiation, which was only about 1/20 of natural radiation. Dr. Teller apparently just doesn't worry as much as biologists do.

'Common Sense' doesn't always make sense

With respect to "common sense" which Professor Teller says assures him not to worry about (0.17 rem) 170 millirads per year, there are a number of terribly important examples in scientific history that he might think about. And if he did think about them, he might lose a little faith in the "common sense" prevalent at various periods in the history both of physics and medicine.

When the evidence was provided by Galileo that the earth revolved around the sun, "common sense" told the experts of that day this must be heresy. But the earth *does* revolve around the sun.

When someone suggested the earth was round, not flat, "common sense" of the times must have suggested the idea was really ludicrous. The earth is not perfectly round, as Dr. Teller knows, but much closer to it than flat.

When Semmelweiss suggested that doctors and midwives were the cause of childbed fever, the Vienna medical authorities knew from "common sense" that Semmelweiss must be wrong. Yet women do not die of this dread disease today because we finally learned that Semmelweiss was right.

When Hammond and Horn proposed in 1954 that heavy cigarette smoking produces an enormous increase in cancer of the lung, "common sense" told the medical community and just about everyone else this must be silly, since obviously so many people smoked and didn't have lung cancer. Essentially no reputable medical authority disputes Hammond and Horn sixteen short years later.

Would Professor Teller consider re-thinking his reliance on "common sense" with respect to radiation hazard now that Dr. Stewart has shown that radiation levels just twice annual natural radiation *doubles* the incidence of childhood cancer and leukemia if the infant receives the radiation in the first 13 weeks of its utero life?

We must attend to Professor Teller's further reassurances in his letter to Senator Gravel. Professor Teller says:

> Procedures exist in many cases by which radioactive body-burden can be removed if this body burden should exceed the "Maximum Permissible Dose." Thus, occasional exposure of people to excessive radiation can be counteracted and the relatively small hazard to a limited number of people need not be incurred. All of this can and should proceed under the present guidelines.

It is only after sober consideration of the deep implications of Professor Teller's statement that the reader's fury will mount. Aside from a few rare types of radioactivity poisoning, there is practically nothing known about how to get radioactivity out of people. Moreover, some of the chemicals used to attempt to remove radioactivity from human beings are exceedingly dangerous, and may be even more dangerous than the radioactivity. But Professor Teller seems utterly oblivious to any of these medical facts. He is willing to let people be exposed to the "maximum dose," and to *hope* some method is available to clean them up once they are poisoned. So, he suggests leaving the allowed amount of radioactive poison as high as it is while hoping that some way might be found to salvage the life of those injured. We wonder how many individuals in the public-at-large appreciate themselves being the victims of this promotional philosophy. It may well turn out, Professor Teller, that the public may prefer not to be poisoned at all!

But then if the "allowable" dose of radiation is lowered, and thus prevents poisoning by radioactivity, Professor Teller would undoubtedly fear for the viability of his favorite program, the Plowshare "peaceful" nuclear explosives program.

In discussing the benefits to be lost if people are better protected against radiation hazards, Professor Teller tells Senator Gravel, in the same letter, the following:

> A second benefit which might suffer unnecessarily by strict

regulations is the Plowshare project. Due to exceedingly great caution, the development of the project has been slow. We have to rely on plans and guesses concerning possible benefits.

What Professor Teller is telling us is that we must explode enough of his "peaceful" nuclear bombs, and thereby spew radioactivity into the biosphere, in order to find out the possible benefits. The general Plowshare philosophy is that nuclear bombs must be good for something!

Professor Teller's last recommendation in the letter to Senator Gravel is remarkable. He is speaking about doses of 0.17 rads—the amount which goes with 32,000 extra cancers per year for the United States. His recommendations are as follows:

> When the effect of released radiation reaches at any instant the average (that is, when the effects of the United States average is doubled) in any human organ, or when there is enough fall-out to threaten such a concentration in a human organ, protective measures should be taken at the expense of the parties responsible for the radioactive contamination. These protective measures may consist in the removal of some radioactivity from the affected people. In case of tritium contamination, methods for doing this are available. In case of strontium or iodine, there are methods to accelerate elimination. With more research and development these methods can probably be improved. In other cases, contaminated materials could be removed from the food chain. These protective measures shall proceed promptly whenever the effected people request it.
>
> It should be realized that at present protective measures are only partially effective. In the important case of tritium they are quite effective.
>
> By making them available one can greatly reduce needless worry. Thus, in the long run one can hope to hold actual damage to a minimum even in the few cases where releases exceeding the national average have taken place. In this way it can be made clear that the result of inadvertent release will become primarily an inconvenience to the affected people. It is improper to concentrate on the frightening aspects of improbable fatalities when there are good prospects that methods for preventing such fatalities can be developed.

Few readers will believe their eyes on reading what Professor

Teller recommends for their future. We certainly can say that with Professor Teller as one of Plowshare's leading friends, Plowshare doesn't need enemies. The Atomic Energy Fairyland has indeed many interesting prospects in store for U.S. citizens. As to Professor Teller's "needless worry"—Who, me?

The Lawrence Radiation Laboratory at Livermore is a strange and wondrous place. We have learned so very much about how nuclear technologists think. One does have to admit that the laboratory is a bit remote, located as it is in the isolation of the Livermore Valley. So, perhaps, nuclear technologists there are at a disadvantage in trying to have their thinking in tune with the rest of humanity.

Blowup

In 1957, Professor E. B. Lewis published a very important paper concerning the production of leukemia by radiation in man. His scholarly considerations led him to suggest that there might be no safe dose of radiation, and that every dose of radiation would produce its proportionate share of human leukemias.[19] More radiation, more leukemias. And Professor Linus Pauling published similar predictions concerning cancer, leukemia, and genetic injury in human beings in proportion to the dose of radiation they received. His calculations were sobering, to say the least, and most disturbing concerning the consequences of irradiating large numbers of human beings.

Anyone who has read Pauling's excellent book, *No More War,* must be impressed with the depth of his considerations and his ability to predict the consequences to human beings of receiving nuclear radiation, at a time when much of the direct human evidence was not yet in. But Professor Pauling has been decades ahead of others in so many important areas of science and medicine that there is really not so much reason for surprise that his radiation predictions were also superb. Regrettably, the world seems determined to stay two decades behind Linus Pauling in his statement that there can be "no more war."

Events brought the issues into sharp focus—the issues of the real hazards of nuclear radiation to man. These events proved to be a turning point of major proportions in our understanding of the real reasons why environmental catastrophe is upon us and why resolution of the crisis will prove so enormously difficult.

Tamplin and his group had essentially completed the broad framework of prediction of where radioactivity could go following nuclear explosive detonations. The worst case expectations were now estimable. And from these data and a reasonable knowledge of how radioactivity of various sorts gets into man, it became possible to predict the radiation dose that he receives thereby. To be sure, minor features of wind patterns could change the amount of radioactivity in a particular location, but broadly the task of predicting how many people would receive what burden of radiation from nuclear detonations was rapidly becoming a completed one. We both turned our attention more and more to the issue of precisely what effects we would anticipate from atomic energy programs such as Plowshare and other AEC programs.

We reviewed the work and predictions of Lewis, Pauling, and others concerning effects such as cancer, leukemia, and genetic injuries. Additionally, an entirely new area of medical investigation, the study of chromosomes, had opened up in the period of 1960. And the chromosomes, upon which the hereditary units, the genes, are distributed, loomed as especially important potential targets for possible irreversible damage by ionizing radiation. Indeed, the developing data concerning radiation sensitivity of these all-important structures appeared quite ominous. We began to ask ourselves a serious question.

Nuclear technologies begin to burgeon

Lewis, Pauling, Schubert, and others, had raised grave questions of injury to human beings from radiation doses associated with the fallout from nuclear weapons testing. And here we were exploring the future of burgeoning technologies such as nuclear excavation of harbors, canals, stimulating natural gas production with nuclear explosives, nuclear reactors for electric power which were to begin ringing every major metropolitan center in the land, and radioisotope shipments which were finding their way into research laboratories, industries, and hospitals in ever increasing amounts. All this was proceeding apace, ready soon to skyrocket in application, under a set of allowable doses prescribed by the Federal Radiation Council Guidelines *many, many* times higher than doses of fallout that had led Pauling and others to grave predictions of human misery and death.

Indeed, the doses to be allowed to the population were some 20 times higher than those which had been worried about in the 1950s—and which had led to Professor Pauling's grave concerns for the welfare of human beings and irreversible pollution of the planet. Why were we not alarmed?

Plowshare plans for bigger and better explosions

Plowshare, the "peaceful" nuclear explosives program, worried us more than anything else except for nuclear weapons. At that moment we hadn't even thought much about nuclear reactors. From rare experimental detonations with small nuclear explosives, the men of this program were whetting their appetites for bigger and more nuclear explosives—commercially! They were cheap, the Plowshare advocates advertised—why we can have thousands of such detonations annually—and just around the corner.

Worse yet, the chief advocate for Plowshare, Professor Edward Teller, called Gofman in for a talk during this period. He explained his view that, at the worst, radioactivity and radiation dosage to people wouldn't kill as many human beings as many other environmental hazards, and many things we accept without thought or concern—like 50,000 people killed per year in automobile accidents.

Gofman told Dr. Teller he didn't know how bad the radiation hazard might be—that, as Dr. Teller knew, we were in the midst of serious investigations of precisely that problem, and in spite of his assurances, we were indeed very concerned. Dr. Teller did not consider the effect of small amounts of radiation to be of consequence to human beings and said he thought we shouldn't be worried about it. Plowshare, he said, could do many wonderful things for man with nuclear explosives and what Plowshare needed was permission to give the public three times as much radiation as was currently allowable under the Federal Radiation Guidelines.

Dr. Teller told Gofman that with his national standing and prestige, he (Gofman) could be of enormous help in prevailing upon authorities to raise the permissible dose limits. And that this would allow a full blooming of the "peaceful" nuclear explosives program. Gofman simply told Dr. Teller he couldn't conceivably be of help in *that* manner. We were very concerned that even the present guideline radiation might mean extremely grave risks for the human species—and with that concern, suggesting three times more was unthinkable. Lest anyone misunderstand, we are certainly prepared to believe Dr. Teller was totally sincere in his unconcern over the radiation hazard. The real problem is that technology bends people into such unconcern and leads them, with utmost sincerity, to disastrous positions on matters of pollution of the environment.

It was a time for worry

At this point we were more acutely worried than ever—the radiation effects problem must become immediately dominant. Here we had pressure from nuclear projects to raise the radiation dose allowed to people three times more—and this meant *sixty* times as high as the dosages Linus Pauling and the 11,201 scientists who signed his petition considered as representing a grave problem. This made us face the problem squarely—why weren't we worried as hell? Why weren't we worried about the prospect or 20 or 60 times as much radiation to the public—an amount likely on the early horizon with the determined and rapid pushing of AEC programs. Had Pauling ever been proved wrong? Had anyone shown that Pauling's dire predictions were overstated? We knew of no disproof of any of Pauling's work, nor of his estimates of radiation hazard.

It can and should be asked, since we couldn't believe in, or defend, the allowable dose of radiation by federal standards, why we didn't attack those radiation standards as being dangerous to human life? Why didn't we sound an alarm about burgeoning Plowshare activities that had hopes of permitting even a three-fold higher dose than existing standards?

Our answer is that we were mesmerized by what represents a fantastic error of thinking that has characterized atomic energy development and just about every other technology capable of releasing by-product poisons upon the public. And understanding the basis for such mesmerization is *the* most crucial single issue facing everyone concerned about reversing the catastrophic downward plunge of the environmental crisis. And it is no accident that such foolish mesmerization is widespread even in those primarily charged with concerning themselves with hazards to the public, like us.

It's dangerous to go ahead if safety is not assured

This error is that if one can't *prove* a particular dose of technological poison (radioactivity in this case) is unsafe, the technology (atomic energy in this case) is allowed to proceed full tilt, even though the harm it may be doing to human life and toward irreversible poisoning of the planet Earth are both extreme.

How does one get in a position of subscribing to such a dangerous Rasputin-like mesmerizing spell? We did, for all too long a time. What we can say is that we, at least, have by now broken completely the bonds of this nonsensical spell, while so many of our AEC colleagues are totally and blissfully still mesmerized.

It pays to examine at even closer range why we failed to realize and to speak out concerning that which should have been obvious to us as early as 1963—indeed, even before we organized the Lawrence laboratory Bio-Medical Program. We realized Pauling and others had never been proven wrong concerning the hazard of radiation to man. We said in 1964 we could not defend the radiation standards set by the Federal Radiation

Council, but we didn't fight them or the atomic energy programs developing under their blanket of protection. It is obviously erroneous public health practice to go ahead when safety is not assured. And safety is not assured when either of two situations exist: (1) When a hazard is *known* to be large under the operating conditions for the technology; and (2) When the hazard is *unknown,* or not fully understood.

Technology promoters ignore even the known hazards

It is bad enough to proceed recklessly with a technology in the face of a serious known hazard; it is incomparably worse to proceed when there are possibly many-fold larger unknown hazards. It is a monumental example of human arrogance that technology promoters, and even those ostensibly charged with public health protection, make pious pronouncements that the hazards are "acceptable" when they themselves realize the hazards are not even known *crudely.* Thus, in the case of atomic radiation, the hazard of development of cancer and leukemia is by now reasonably accurately known; the genetic hazard is not well known, but the best estimates indicate it is likely to be 5 to 50 times as large an effect—even potentially an irreversible disaster for the human species. So the uncertain hazard, or unknown hazard, should lead to considerably more caution than the serious known ones. Does it, in practice, lead to such caution? Absolutely not!

Once we became sufficiently awake to realize the frightening possibilities of the unknown hazards, we pointed this out over and over again to various of our atomic energy colleagues in the Washington office of AEC Biology and Medicine. When we asked the crucial questions mildly, we got benign smiles, a pat on the head, and the vacuous statement that there was a great deal of research to be done on such important unsolved problems. But when we pressed the matter further and asked, "Why in the world are we proceeding with developments industrially throughout the land, potentially preparing to expose the whole public to such unknown hazards?" the reaction was more grim and unfriendly. "Hell," we were told, "it will take 20 years to prove or disprove the problems you're bringing up. Atomic energy can't wait that long. Do you want to simply stop progress?"

The meaning of that as an answer was not lost upon us. Instead of realizing that our ignorance indeed demanded that atomic energy go slow, the atomic energy promoters, including "the public health protectors" among them, said full speed ahead, visualizing themselves as modern champions of the pioneer spirit.

But it *is* important to realize that few human beings consciously can suffer themselves to commit evil deeds. Atomic energy promoters do not consider their actions evil, or their intentions evil. Nor do we consider the intentions evil either. The only point is we understand their actions are the result of layer upon layer of soothing rationalizations, and we understand only too well the futility of working through all such layered rationalizations to expose the confrontation such men themselves cannot face.

The mythologies of rationalization

In the atomic energy field, but by no means limited to it, promoters have carefully cultivated what we would call mythologies of rationalization. Consciously or unconsciously developed, these mythologies can most properly be regarded as an invention to serve the purposes of a promoter of technology. The mythology must provide some hope that the pessimistic hazards will somehow disappear or never appear. If this hope is provided, it is the most elementary step in the world for the promoter, even the "protector," to gradually translate, by rationalization, the hope into a belief and the belief into fact. And once this is accomplished, there are no longer any worries about hazards.

Let us examine two of the outstanding mythologies so ingrained in the atomic energy fairyland as to have lulled even us to sleep for all too long.

The first such myth goes as follows: "Most cases of radiation-cancer, leukemia, or other injury have been observed when the dose of radiation was quite high." This was true up until recently. Therefore, the myth continues, *maybe* there exists some smaller amount of radiation that is a "safe threshold." By this is meant that possibly cancer, leukemia, or the even more deadly genetic injury won't occur provided the total radiation dose is kept below some magical number. Certainly this is a most convenient hope, especially for atomic energy promoters, but it is a hope *unsupported by any* scientific evidence. Numerous supposed items of evidence which have been brought forward to support the hope of a "safe threshold" have been discredited by numerous reputable scientists repeatedly. They were discredited so thoroughly that the highly responsible International Commission on Radiological Protection simply refused to consider the possible existence of safe radiation thresholds in its own evaluation of the numbers of cancers and leukemias to be expected for each amount of radiation exposure. Even the Federal Radiation Council refused to consider acceptable any evidence for a "safe threshold" of radiation.

But the promoters of atomic energy technology keep hoping that somehow, somewhere, evidence will be developed that a "safe" amount of radiation exists. At Lawrence laboratory, the reins of leadership of the Bio-Medical Program passed from Gofman's hands in 1966 so he could return to laboratory research. The present leadership (Roger Batzel, associate director, and Bernard Shore, division leader) reflects the atomic energy promotional philosophy. How? We are treated to the spectacle of the Bio-Medical Division now listing as one of its major endeavors a program entitled, "The Search For a Safe Threshold of Radiation." Seek and ye shall find—even the little man who isn't there.

In the early period of atomic energy, the direct evidence of injury, both for man and experimental animals, was for large radiation doses, 100 rad units or more. This provided a fabulous opportunity for the promoter. Maybe there is a threshold somewhere below 100 rads. And if we promoters never give anyone over, say, 20 rads, we'll be below the safe threshold—and then we can go forward with exploitation of the technology—gung ho! Having translated hope into reality in their minds, the promoters proudly proclaim their responsible concern for the public health —they would never consider giving the public more than the (hope-to-reality) safe threshold dose. No. No.

But time has a way of passing and providing evidence that shatters the promoter's dream hope. Both in men and in experimental animals strong direct evidence has shattered the promoter's dream of a safe threshold with respect to cancer and leukemia production by radiation. In the experimental animals direct evidence clearly shows production of cancer down in the 10-20 rad dose region. What happened to the hoped-for "safe" threshold?

Dr. Stewart's important contribution

In the human the demolition of the "safe" threshold idea for radiation is infinitely more devastating. We owe this contribution to humanity to the works of Dr. Alice Stewart and her collaborators in England. Her studies, first published in 1956, deal with the effect of diagnostic radiation of pregnant women, which, of course, also means radiation of the fetus in utero. The amount of radiation (of x-rays no different from the ionizing radiation of atomic energy activities) to the infant in such examination is only about 2 rads per average examination. And Dr. Stewart found that children who had been radiated in utero had a 50% increase in the number of cases of cancer in various forms, as well as leukemias, during the first 10 years of life.[20] This represents an enormous sensitivity to radiation—far more serious than even the pessimists had anticipated.

Considering the shattering impact of Dr. Stewart's work, it is not at all surprising that the atomic energy promoters and the radiologists scoffed at the findings, dismissed them. But Dr. Stewart persisted in her important researches, abundantly confirmed her earlier findings. Not long after, Professor Brian MacMahon confirmed her work in the United States.[21] And most recently, Dr. Stewart has presented even more damning evidence.

By careful study of thousands of cases of childhood cancer and leukemia and counting the number of x-ray pictures taken during the pregnancy, she has found that 1½ rads *doubles* the frequency of childhood cancer or leukemia if the pictures were taken in the latter half of pregnancy. If taken during the first three months of the pregnancy, only ⅓ of a rad is required to double the frequency of cancer or leukemia in the offspring during the first 10 years of life.[22] And ⅓ of a rad is a long way from the 100 rads the AEC promoters are so fond of speaking of *still* as the lowest dose where evidence of cancer is seen in human beings due to radiation.

For those who wonder why the Atomic Energy Commission promoters are so refractory to allowing this important new knowledge to diffuse into their brains, it requires the reminder that shattering one's fondest hopes comes hard, no matter what the truth be. The idea that a safe threshold of radiation *must* exist, even if it doesn't, appears glued into the brains of atomic energy promoters with the strength of epoxy glue—and perhaps with the known strength of epoxy, such promoters fear removal of this "fond hope" may take too much brain with it.

Prof. Radford attacks the 'safe' threshold concept

The reader may find it incredible that the refractoriness of the AEC and its supporters in the Congressional Joint Committee on Atomic Energy could be so extreme. Examine the record. When Gofman and Tamplin presented evidence to the Senate Subcommittee on Air and Water Pollution—evidence that "safe" threshold concepts had no support in scientific evidence—the Atomic Energy Commission promptly replied to Senator Edmund S. Muskie that Gofman and Tamplin were wrong because they had failed to take into account that a threshold existed below which radiation dosage could *not* produce cancer. The heady drink of a $2.5 billion budget still was making the AEC staff see the little man who wasn't there.

Soon thereafter, at hearings of the Joint Committee on Atomic Energy, Professor Edward Radford of Johns Hopkins University expressed his indignation that the AEC should seek shelter in the "safe threshold" concept when the International Commission *and* the Federal Radiation Council both have rejected it as being an unprovable concept with respect to protection of the public health. Dr. Radford told the Joint Committee that the AEC's seeking shelter behind the so-called "safe threshhold" meant a *reversal* of everything sound biologically, with respect to such envionmental pollutants as radiation-emitting substances, that had ever been won over a period of years.[23] No doubt the vast majority of knowledgable biological scientists with a concern for humanity would have seconded Professor Radford's comments vigorously.

Congressman Chet Holifield, chairman of the Joint Committee, knew precisely what Professor Radford was saying. The AEC's duplicity was on the record. And Chairman Holifield said explicitly in a measured voice, "The AEC *will* follow the ground rules set down by the FRC and ICRP"—the conservative ground rules meaning no reliance on "little men who may be there"—a "safe" threshold.

The AEC does an about-face

The Atomic Energy Commission was on the spot—their staff had made themselves ridiculous in the unseemly haste to criticize Gofman and Tamplin, and had thereby raised a storm over an issue they could not possibly defend, ever! So they promptly reversed themselves, saying, "We are not aware of any standards of radiation exposure that had ever been set assuming a 'safe' amount of radiation." The words were slightly changed—but they were now in effect, eating their previous words in print, because they were caught with both hands in the cookie jar!

And on March 4, 1970, in the city council chambers of New York, Councilman Theodore Weiss looked squarely at Dr. William Burr, Associate Director of the AEC Division of Biology and Medicine, and asked, "Do you assume a safe threshold of radiation for cancer and leukemia production?" And Dr. Burr said, "I know of no standard-setting body that assumes a safe threshold of radiation for cancer production."[24] Dr. Burr was now saying, in effect, every amount of radiation should be considered to produce its proportionate share of cancers and leukemia. How very interesting: Caught in a vise between the Joint Committee on Atomic Energy and Professor Radford's direct challenge, the AEC had reversed its field completely!

But the promoter never forgets what is best for promotion and the AEC repentance was quite short. Losing the crutch of a safe threshold that could be used for false reassurance of an anxious public seems like too big a loss for AEC. And thus, two short weeks after Dr. Burr said we know of no one counting on a safe threshold, Dr. Glenn Seaborg, Chairman of the AEC, appeared on the CBS Morning News television show. Joseph Benti, CBS newsman, asked, "Do you consider there is any safe threshold amount of radiation?" and Dr. Seaborg replied, "We believe there *is* a low threshold." Flip! Flop! Flip! This time Professor Radford wasn't watching, and the statement slipped by unchallenged. How long is the American public expected to tolerate this behavior on such an important issue?

For public health purposes there is only one way to deal with a question like, "Is there any safe threshold amount of radiation below which cancer doesn't occur?" Either we know positively, and beyond any reasonable doubt, that such a threshold exists, and then we can relax about doses *below* that level; or we don't know anything at all about such a possible safe threshold. Suppose we have such a situation. Thus, suppose under a circumstance of no such threshold, one calculates 32,000 extra persons will die annually of cancer and leukemia if exposed to a particular dose of radiation. On the other hand, if a threshold exists above some particular dose of radiation, there would be 0 extra similar deaths annually. So the decision lies between 0 and 32,000 deaths, and let us suppose nothing allows us enough insight to choose between these two extremes or any number in between. Which number should we assume for planning purposes in a technological enterprise?

Historically and up to today, promoters like the Atomic

Energy Commission assume the zero value, or something very close to it. Not because of malevolence, but because they hope it may be correct; certainly it is the convenient choice which interferes least with selling the technology to the public. But under the circumstances described, choosing zero and acting upon that choice represents irresponsibility of the highest order in neglect of the public health and safety.

The public should have right to decide

The only reasonable choice, in the absence of a firm answer, is to use the 32,000 calculated deaths in planning whether the technology should even be considered. Pauling realized this clearly. For reasons we shall never understand, we saw these same possibilities, said that the risk was between 0 and 32,000 and we took no steps to sound an alarm based on the higher number. It is so obvious to us now that we should have done so. And for all future technologies, it is imperative that a similar error not be allowed to occur if a habitable earth is to be preserved. And the more persistent the poison is in the environment (and radioactivity persists for hundreds or thousands of years), the more crucial is this decision for humanity. To proceed in any other way must be described for what it truly is. One apt description is the playing of "Russian roulette" with the health of the public and the future of the human species on earth.

Another description is that a value judgment has been reached, namely, that the benefits of the technology are worth the annual murder of 32,000 human beings. Should the promoters of a technology be the ones to make this important decision? They *think* so, in a supreme demonstration of arrogance and indifference. From them come such words as "the problem must be left to the experts," or "the public will be alarmed," or "the public can't be relied upon to understand." Surely the public can understand the numbers 0 and 32,000, and if told the truth, will emphatically say "No, thanks" to the promoters' dream project. And *this*, not fear of public inability to decide, is precisely why the technological promoters, atomic promoters foremost among them, don't want the problem put in the hands of the public, where it belongs.

There exists a favorite game of self-deception engaged in by promoters of technology. The game guarantees, by a self-fulfilling prophecy, that the technology will not be thwarted. The deception is particularly gratifying to the so-called "scientific" supporters of the technology—men who hold excellent grants of large sums of money to do research on health aspects of radiation. And this game goes as follows: Let us suppose we expose people to a poison such as radioactivity, sufficient to produce one death each year out of every thousand people so exposed. This is a death rate that can be regarded as a calamity of unparalleled proportions—10 times as big as the entire leukemia incidence in the United States. Now, if we expose 100 people to this amount of radioactivity, what do we expect? If there's one death expected in a thousand, that's only 0.1 of a case in a sample of 100 people. But one can't observe 0.1 of a human case. Either we observe *no* cases or one case, two cases, and rarely more than two. If we study 10 separate groups of 100 people, the most likely outcome is to find one case developing out of all 10 groups of 100 people each. This is true because 10 groups of 100 people add up to 1000 people, and we have already said we're expecting *one* case in 1000.

The key issue, which must be understood by every intelligent person, and is so easy to understand by anyone, is simply this: *If* we study only *one* sample group of 100 people, nine times out of ten we won't see that one death due to the poison. And then, lo and behold, the scientific apologist for the technology promoter piously announces:

"The suspected poison is harmless at this dose—no injury was observed." Hogwash.

"This dose of poison must be below a 'safe' threshold." Eminent hogwash.

"The effect is so small that it can't be detected except by elaborate statistics." More hogwash!

"The effect is so small it is negligible and is readily drowned out by more important factors." Absolutely not. NO. NO. NO.

All such ludicrous statements have only one result, intended or unintended—allaying legitimate concern over a massive threat to health from a by-product of technology. There is no need to question at all the motives of the scientific apologists, or to suggest they are being deliberately and immorally deceptive. No, the entire structure of thinking in evaluation of poisons, or potential poisons, has simply been erroneous—and deeply imbedded into the scheme of things. And unless we get out of this morass, the environment and everything alive in it are headed for trouble of the gravest kind.

In the study cited above, the answer is very obvious. If we study too few people, we are bound to fail to observe the correct answer. Do the wrong experiment, and most certainly the wrong result will be obtained! Wrong experiments of this sort—studying too few people or too few experimental animals—are the cornerstone of all the erroneous, platitudinous reassurances we hear from atomic energy promoters. Worse yet, they don't even realize the experiments were the wrong ones.

Repeating the wrong experiments over and over

The Division of Biology and Medicine of the AEC has spent a billion dollars ostensibly to determine such answers as the risk of cancer and leukemia from exposure to ionizing radiation. Much of the research is of very high quality—but over and over again it is the wrong experiment that has been done. Not once, but over and over again. One can find mountains of published reports on experimental animals given a dose of radiation which produces cancer in 50% to 100% of the radiated animals. Of what possible interest are such experiments? Of what possible relevance to the real problem of hazard of radiation exposure? If one thinks for a moment, it is plain that radiation doses that provoke cancer in 50% of exposed subjects in a brief time period are *obviously* enough to wipe out the species. For such profound conclusions, we don't need to waste precious time, money, and effort.

But of experiments that can properly answer the question of production of cancer in one out of 1,000 animals, which corresponds to tragedy of unparalleled proportions in man, the experiments are virtually non-existent. What answers does one get in asking why such experiments are not accomplished while totally irrelevant ones abound? "That would be too difficult." "That would take too many animals." Would it? A quick calculation from evidence obtained on groups of animals given high doses of radiation indicates the study of 1,000 rats, or a few thousand rats, would provide ample information. Considering the dollars expended and the fact that rats are not in short supply, the proper experiments should have been accomplished long ago. Fortunately the message has *begun* to sink in, and better experiments are now being done.

But atomic energy is not alone in its attention to the wrong kind of experiments. Evaluation of practically every potential toxic substance that is a by-product of technology has been equally poor—equally devoid of understanding of the real nature of the problem. Dr. Saffiotti, Scientific Director for Carcinogenesis at the National Cancer Institute has recently gone right to the heart of this matter in his criticism of studies of cancer-producing properties of toxic chemicals of commerce and industry where the scientific investigators set up experiments looking for 2% or 20% of animals to get cancer, and then label the poison as *safe* when they don't find 2% or 20% cancer. They should be looking for 1 in 1,000 or 1 in 10,000 animals to get cancer if they really want to know something about the poison—that is, something truly relevant for health protection.

The study of too few cases may spell disaster

In the human being, the situation is even more outrageous— and overtly deceptive. We have already alluded to the study of 100 people where 0 cases will be expected 9 times out of 10, and if 0 cases are observed in one sample of 100 people, the pronouncement of "safe " is made, when indeed disaster may be the correct answer. Obviously it *may* require 10,000 persons observed, or 100,000 persons observed, to discover directly that 1 out of 1,000 are developing cancer as a result of a specified dose of radiation. Does this justify a pronouncement of safety because of a totally inadequate set of observations of 50 or 100 persons?

Fortunately, we have not yet reached the Nazi-principles of human experimentation under which we would specifically design experiments to radiate 10,000 or 100,000 persons just to get this answer. We fervently hope that point will not ever be reached. But it is even more cruel and more inhumane to pronounce "safety" from the totally inadequate study of 50 or 100 persons, and from that to go ahead with an allowable exposure of 200,000,000 people at such a radiation level. *This* represents an experiment in inhumanity of the greatest arrogance and disdain—and to add insult to injury—once the whole 200,000,-000 people *are* exposed, it becomes impossible ever to learn the effect of the radiation upon them because there is no unexposed group left for comparison.

The credulity of the average thinking citizen might be strained in trying to believe that such a massively inept approach characterizes health and safety considerations in atomic energy. An illustration will help. After presenting our evidence that 32,000 extra cancers and leukemias could result from the currently allowable radiation levels to Senator Muskie's Subcommittee on Air and Water Pollution,[25] we were invited over to the Joint Committee on Atomic Energy for some discussions of the radiation hazard question. Chairman Holifield of the JCAE helped demonstrate for us how widespread and pernicious erroneous thinking can be in leading to potentially disastrous public health results. In this case, even the chairman of the supposed watchdog Congressional Committee appears to have been thoroughly taken in by the inappropriate kind of evidence concerning hazard.

"How can you two sit here and tell us that the currently allowable amount of radiation can finally lead to 32,000 extra cases of cancer and leukemia per year? How can you tell us that *anyone* would be hurt when we *know* that the standards were set as follows: First, the injury level was determined; then we went 10 times below that; And then 10 times below that—and the standards were set there. If we're 100-fold below the injury level, how do you two come up with predictions of 32,000 extra deaths per year?" Certainly a fair question by Chairman Holifield.

Chairman Holifield is fed an old chestnut

We knew how we got our numbers. It was very simple. We had studied the direct results of how many cancers and leukemias had already occurred in people either exposed to radiation from atom bombs (Hiroshima-Nagasaki) or from medical exposure. Any fourth grader could take *that* evidence and arrive at the same numbers we did. Even the AEC could, if it would but try —it isn't that the calculation is hard.

So we said, properly, "Congressman Holifield, you have been misinformed. We don't know precisely how you got the wrong information about cancer risk, but we do know it's wrong."[26]

On leaving, Gofman said to Tamplin, "Where in the world did Congressman Holifield get that rubbish about the standards being set 100 times below the danger point? What danger point? Whoever proved one?" "Beats me," said Tamplin, "you and I have spent a few years poring over the evidence—it couldn't possibly have escaped us." So we resolved to go right back to California and find out what mythology had been fed to the congressman who chairs the Joint Committee on Atomic Energy.

It didn't take long. Over a period of years the JCAE had held extensive hearings on the general subject of radiation hazards— hearings containing elegant testimony that should clearly have dispelled confusion in the congressman's mind. Our copies of those hearings were already worn thin by our previous perusal. We must look; we must find out who misled Chairman Holifield. Just as we suspected, the old story of *supposed* safe thresholds of radiation was the culprit.

Years ago a number of workers had been exposed to radium

salts as a result of working in the dial painting industry on luminous dial watches. After a period of years such people started to die of bone cancers as a result of radium radiation poisoning. For those exposed to high amounts of radium, the death rate was appalling. Some workers, small numbers of them, had been exposed to very small amounts of the radium poison. By simple proportional calculation for one of these groups, a handful of people (50 or so) one estimates that 0.01 of a case of bone cancer is expected.

Obviously, with so few people, you can't observe 1/100 of a cancer case; human beings just aren't packaged that way. And, as related above, most of the times 50 people are studied with such expected numbers, 0 cases will be observed. Instead of realizing that this proves nothing about radiation hazard, precisely the wrong interpretation was made—namely, that this amount of radiation was harmless—that this was below the "magic safe threshold." Utter nonsense! Many great scientists, including Archer, Morgan, Parker, Hems,* had all rejected the radium-dial painter story as being at all indicative of any safe amount of radiation. They said so, right in Chairman Holifield's Radiation Hearings. In print! The International Commission on Radiological Protection rejected it! The FRC, including its chairman rejects it—again and again. But, oh how attractive such *non-evidence* is to the atomic energy promoter seeking any safe port in a storm! Chairman Holifield is not to be criticized for having been taken in by such purported evidence for safe thresholds of radiation. Chairman Holifield admits he is a protagonist in favor of atomic energy—and try as he may to escape it, the dynamic of hope seeking to see only beauty in his dream-child shuts out the seamier aspects.

And so we did indeed understand Chairman Holifield's remarks to us, and we wrote him the following letter, which he published:

*Victor E. Archer, M.D.; Karl Z. Morgan, Director, Health Physics Division, Oak Ridge National Laboratory; H. M. Parker, Health Physicist, Battelle Northwest Laboratories; G. Hems, Department of Public Health and Social Medicine, Foresthill, Aberdeen, Scotland.

UNIVERSITY OF CALIFORNIA,
LAWRENCE RADIATION LABORATORY,
Livermore, Calif., December 1, 1969.

HON. CHET HOLIFIELD,
Chairman, Joint Committee on Atomic Energy,
U.S. Capitol, Washington, D.C.

DEAR CONGRESSMAN HOLIFIELD: Both of us were deeply honored by the opportunity of some two hours of frank and substantive discussion with you and your colleagues last week. Especially is this so because both of us are intense admirers of the devoted and untiring efforts of the Joint Committee on Atomic Energy to bring to light all the true facts concerning radiation hazards. The various Hearings you have held are unequalled as a monumental contribution to the public welfare and health.

In our discussion you asked us a very specific question, "How can you tell us there is a potential hazard at certain dosages when we have been assured that the hazard level is approximately 100-fold higher?

We answered, "Congressman Holifield, we believe you have been misinformed."

We know that you needed more answer than that. Based upon the evidence and calculations, we knew that what we were saying had to be true, but we did not know *how* it had come about that deep *mis*-information had come to the Joint Committee. We resolved, therefore, to go right home and find out how this had, indeed, come about. After careful study of many of the Hearings of the Joint Committee on Atomic Energy, we believe we have complete understanding of the *specific nature of the misinformation.*

It is the purpose of this letter and the attachments to explain all of this to you. And we are prepared to defend our analysis of this situation in any format the Joint Committee would find helpful. We believe, however, our analysis will speak for itself.

Specifically, we refer to the Radium Dial Painter studies reported to you by Dr. Robert Hasterlik at the Hearings (87th Congress, Part *1*, p. 325) and by Dr. Robley Evans at the Hearings (90th Congress, Part *1*, p. 265).

Dr. Hasterlik interpreted his findings correctly when questioned by Congressman Price. Dr. Evans, in our opinion, grossly *misinterpreted* his own data, but undoubtedly with total sincerity of purpose.

Our analysis attached shows both sets of data consistent with each other. In *striking contrast* with Dr. Evans' claim that the data indicate a *threshold* of radiation below which cancer doesn't occur, our analysis indicates *nothing of the sort*.

1. Neither the Hasterlik data nor the Evans data *can even remotely* be construed to suggest any safe "threshold" below which cancer doesn't occur.

2. The data from both researchers are perfectly consistent with cancer production right down to very low doses, and this could very well be a linear relationship over much of the entire dose range from low doses upward.

We are both dismayed that the Editorial Board of the "British Journal of Radiology" and the Editorial Board of "Health Physics" did not catch the indefensible claim of Dr. Evans that a threshold exists.

Worse yet, we are dismayed, indeed, by Dr. Evans statement that his "proof" of a threshold is the *cornerstone* of all radiation protection standards. If this be true, then there is little wonder that the cornerstone of radiation protection standards is made of quicksilver.

We believe, after careful study of this particular fiasco, you may be more understanding of our total lack of confidence in the underlying basis for existing radiation standards. However, we are certain everyone concerned in informing you was well-intentioned.

Since we know this information will be of great interest to the AEC, we feel you will approve of our sending copies of this letter and the enclosures to Chairman Seaborg and Dr. John Totter.

Assuring you of our deepest commitment to constructive assistance to you in your gravely important responsibilities, we are
Sincerely yours,

JOHN W. GOFMAN,
ARTHUR R. TAMPLIN.

Thus, we learned further the strange phenomena one must necessarily cope with in dealing with the environmental crisis. The human mind has many mysterious compartments; at the right time, it shows a remarkable ability to close out that which it wants so desperately not to believe is there.

If this were all, one might hope to break through the stone wall. But there's more. An even better game is played by the atomic energy and other technological promoters. It's called,

"Poisoning you slowly isn't nearly as bad as poisoning you fast." What's more, there *is* some basis for the credibility of this story, provided you can sift fact from fancy. Long ago, in the early history of radiation and radioactivity, it was found that a certain dose of radiation would produce reddening of the skin. But the same dose, divided into two applications, did not. A higher total dose was required if split into two portions. And in producing acute radiation sickness, which occurs from massive overdose of radiation, it is perfectly true that spreading the same dose over a prolonged period affords protection against such radiation sickness. But one must ask the real meaning of such phenomena.

No proof that 'repaired' body cells won't become cancerous

Obviously the human organism tolerates more radiation if spread over time than acutely—for the production of acute radiation sickness. Why does it? Does it mean that body cells repair the damage wrought by radiation? Everyone knows that acute radiation sickness comes from a loss of certain kinds of cells. Part of the sickness comes from ulceration along the gastrointestinal tract; and part of it from the loss of blood elements, white cells, platelets, and red cells. Given time between application of radiation doses, the remaining unkilled cells can replenish the lining of the gastrointestinal tract, provide platelets to prevent hemorrhage, and white and red blood cells for such needs as fighting infection and transporting oxygen. If the radiation is given all at once, death ensues simply because replacement of cells cannot keep pace with cell loss.

But apparently little understood by so many atomic energy promoters is that this *apparent* "repair" of the body is a "repair" in the sense of providing replacement cells, not "repair" in the sense that irreversible changes in the body cells which will lead to leukemia and cancer 5, 10, 15 years later have not occurred. There is no reason at all to believe that the ability to replace cells in an ulcerated intestine means that repair of cancer-producing damage has occurred. So, when atomic energy promoters point to the fact that slow delivery, or divided delivery of radiation protects against acute radiation sickness, they should understand that this has absolutely nothing to do with late-appearing effects such as leukemia and cancer. Nothing at all! But hope for such a delightful phenomenon quickly overtakes scientific reasoning—and the hope is quickly translated into "evidence."

A radiation-killed cell cannot produce cancer. Provided it can be replaced, as for example, in the intestine, proper function can be restored. But the very cell which provides the replacement can itself be mortally injured—in the sense that it can become the precursor of a fatal cancer 5, 10, 15, 20 years later—and this is what the real concern about radiation is all about, not acute radiation sickness.

Evidence must replace mere hope

It is known that broken chromosomes can undergo mending. Further, it is known that certain damage to the hereditary material, known as DNA, can be repaired, at least partially. Whether these "repairs" have anything whatever to do with the type of injury leading to cancer is totally unknown. There is a proper way to find out whether any "repair" of cells with respect to cancer-producing properties can ever occur over a period of time. It is indeed an important question. We have found *no* evidence whatever that this kind of repair is possible. We *do not* say it is *impossible*. We say no evidence whatsoever has been provided to convince anyone that such repair is possible. Hope means nothing. *Evidence* means everything. We have searched hard for evidence that the radiation damage of human or other animal cells which leads to cancer can be "repaired." We find none.

Counterfeit evidence is abundant, and, unfortunately, is widely quoted and ballyhooed by promoters of the technology. One special example of such counterfeit evidence is readily explained. In certain animals cancer induction by radiation has been carefully studied. What is found is that at an early age a given dose of radiation produces many more cancers than does the same dose of radiation applied later in life. Why sensitivity decreases with age we don't know. It's just true! And we saw the same phenomenon in human beings, where the fetus in utero at less than 3 months is about 150 times as sensitive as the adult with respect to cancer or leukemia production by radiation.

The need for a new set of standards

Clearly, if we irradiate some animals early in life, we'll get a certain number of cancers. If we take similar animals and start radiating them at the same age but deliver the same total of radiation slowly into an advanced age, we get fewer cancers simply because part of the radiation is being delivered later in life when sensitivity to cancer production is dropping steadily. So, when a comparison is made between cancers produced by radiation all at once early in life versus the same total amount of radiation extended into later life, obviously more cancers are found for the radiation delivered all at once—simply because young animals are more sensitive than older animals.

This is no evidence, by the wildest imagination stretch, of the body's ability to repair cancer damage. None whatever. Indeed, had the investigators of this phenomenon done the one experiment they left out, namely, delivering the all-at-once radiation *late* in life, they would have obviously discovered that slow delivery of radiation produces more cancer than all-at-once radiation. But this result would terrify the promoters of the technology —removing a hoped-for weak reed to lean on. Is it an accident that this experiment is not done? An oversight? Or is it an expression of a subconscious desire to favor the technology? We prefer not to try to answer this question. Why dispute motives when they are so complicated?

The key lesson is that whatever evidence has been claimed for leukemogenic or carcinogenic repair is counterfeit evidence. And no other evidence is available. This lesson has not been lost on responsible biologists. The International Commission on Radiological Protection refuses to count on any "repair" of carcinogenic damage; instead they assume all the radiation produces its proportionate share of risk of leukemia and cancer. So does the FRC, on paper. The chairman of the FRC, in calling for a sweeping review of all radiation standards, has made it abundantly clear he counts on no repair concerning cancer or leukemia risk—he asks for standards to be set considering all radiation effects to be cumulative, delivered *slowly or rapidly*. And this is the only way of proceeding sensibly.

Just as with the issue of "safe" thresholds, so with hoped-for "repair." It represents utter and complete public health irresponsibility to count on hoped-for, unknown, prayed-for ways of extricating the technology from admission of a serious hazard of its by-product poisons.

One may pause here and reflect a moment on academic questions versus public health questions. Is it permissible to consider the possibility of there being some radiation dose below which cancer does not occur—the so-called "safe threshold" concept? Of course. Is it permissible to study the possibility that the changes wrought in cells by ionizing radiation, which later result in leukemia or cancer *might* be partially repairable damage? Of course.

The public health comes first

Not only is it permissible but essential to study these questions, as well as many, many others concerning how radiation affects man. Such academic pursuits are of the highest interest—and ultimately may benefit human beings in ways currently unappreciated. But it is of even greater importance to realize that the existence of a question is not the equivalent of existence of an *answer*. Academic questions take time and effort to solve; the public health cannot wait. The public health cannot be compromised by hopes that something will appear through research to mitigate a hazard.

No, we must take the conservative, sound approach that takes the evidence as it is, and we must proceed, not with Russian roulette, but with sensible humane caution in new technological areas. Erring on the side of caution doesn't kill or maim human beings. Throwing caution to the winds, a promoter's specialty, sounds bold—but kills human beings! Madison Avenue can paint a sexy picture of technology; Main Street needs a less flamboyant approach.

Repression and reprisal

By 1967 we had become thoroughly convinced that the entire approach to the handling of public health and safety aspects of nuclear energy development was erroneous. Such fundamental errors as the requirement that the public prove atomic energy projects unsafe rather than that the atomic energy proponents prove the technology safe troubled us deeply. The growing resentment of our bio-medical efforts by Plowshare enthusiasts was especially worrisome, for this technology appeared to us, by that time, to be one of the worst conceivable, blundering concepts dreamed up by man. How to view discriminate and indiscriminate spewing of radioactivity into the environment as a *reasonable* atomic energy project we can simply no longer understand.

This program of digging canals, harbors, making mountain cuts, diverting rivers by blowing up nuclear explosives, and thereby freely distributing long-lived radioactive materials into the environment cannot be done unless one has already made a value judgment that such projects can be equated in value with a certain number of human lives, present and future, and an irreversible contribution to pollution of the earth. The suggestion to Gofman of trying to raise the allowable radiation dose three-fold is itself an indication that Plowshare programs are clearly predicated on the idea that fairly sizeable human exposure will be required.

Recently, an AEC spokesman (unnamed) was asked when the current radiation exposure guidelines federally allowed might be reached (the AEC generally claims they had no intention of every reaching these exposure levels). The spokesman "declined to predict when the limit might be reached, noting the uncertainties of the Plowshare program among other things."[27]

It would be hard to believe peaceful nuclear explosives can be used in any quantity without delivering appreciable radiation doses to man and other members of the biosphere. We have come to regard this program of peaceful nuclear explosives as pernicious in every respect. It seems difficult to imagine benefits to society that would require such a program.

Needless to say, the Plowshare Program enthusiasts, above all, are desperately furious about the thought of not being able to carry their program through with the irradiation of human beings. Not maliciously, but simply because all of their projects ultimately require such irradiation. We cannot subscribe to the idea that because nuclear bombs are cheap, they have to be good for something. We cannot subscribe to Edward Teller's statement to Senator Gravel that if we can't explode enough of the bombs, we'll never know what benefits they can provide mankind. We certainly *can* know and *do* know about the abundant harm they can do.

A deadly way to increase natural gas

While every project of the peaceful nuclear explosives program of AEC and the Lawrence Radiation Laboratory holds pernicious potentialities, most pernicious of all at the present time is that which proposes to stimulate natural gas production by underground explosion of nuclear bombs in gas-bearing regions. Preliminary experiments have demonstrated already that increased natural gas flow can be achieved through such underground explosions. The commercial appetite is whetted by this, and now we hear rumblings of a future with thousands of such underground nuclear explosions to provide us abundant amounts of natural gas—especially because the nuclear bombs themselves are now cheap (courtesy of American taxpayers).

One little problem has reared its ugly head, namely, that the natural gas recovered so far has been radioactive, either due to tritium, a radionuclide by-product of hydrogen bombs, or krypton-85, a radioactive by-product of the "dirty" fission bombs. The Plowshare advocates suggest that the choice of hydrogen bombs versus dirty fission bombs rests upon which one makes the recovered natural gas less radioactive.

When faced with the question concerning the desirability of piping radioactive gas into the homes of unsuspecting public consumers, and thereby irradiating the consumers, who are buying this poison gas, the Plowshare advocates have a ready answer: "We won't deliver the gas into the homes if it is *too* radioactive." And there's the rub. They will dilute the gas with non-radioactive gas to such an extent that no one using such mixed radioactive gas in the home will exceed the Federal Radiation Council guideline dosages of radiation. What this translates into is, "We won't produce any more cancers, leukemias, and genetic deformities than we are legally permitted to." How utterly generous!

If the public clamor, in shock and disbelief, is too great, the Plowshare proponents of natural gas stimulation are prepared to be even more generous—they will dilute the radioactive gas still further with non-radioactive gas. This also requires translation. At one dilution level, let us suppose piping radioactive gas into the homes of a city of 1,000,000 people results in, say, 100 extra cases of cancer plus leukemia per year. If the inhabitants object (that is, if they've been warned to object), then the radioactive gas can be split by dilution with equal amounts of non-radioactive gas and distributed to two separate cities, each of 1,000,000 people. Thereby, the number of cases of cancer plus leukemia in each of the two cities would only be 50 cases. The total number of cancers would still be 100, just as before, but the *indignation* in each city would be reduced in half.

The peaceful nuclear bomb enthusiasts are quick to point out that this is the old problem of benefits versus risks. Beautiful! But whose benefit? And whose risk? The Austral Oil Company recently joined the AEC in one of these nuclear blast gas projects in the State of Colorado, looking forward to early commercial exploitation via thousands of underground bomb detonations.

Gofman recently suggested to AEC Commissioner Larsen

that it appeared the radioactive gas stimulation project of Plowshare had a strange type of benefit-risk calculation in it. The Austral Oil Company will sell unsuspecting customers contaminated radioactive natural gas and thereby derive a monetary profit. This is the benefit side. The consumer will breathe the contaminated gas and a certain proportion of the consumers will thereby contract leukemia, cancer, and have children born with genetic deformities and genetically determined diseases. This is the risk side of the equation. How do we balance these?

Commissioner Larsen felt this might be an extreme view on Gofman's part, and he suggested a "more reasonable" view. Commissioner Larsen suggested thinking of the Austral Oil Company as the "vehicle" by which this wonderful new technology was being brought to the American public. Only one thought occurs to us upon listening to this "vehicle" story—if that's a vehicle, it really is too bad they ever invented the wheel!

Some of the more sober Plowshare enthusiasts for natural gas stimulation by nuclear bombs speak quite rationally. They say "Look, the natural gas reserves *are* dwindling, and we do, as a country, need additional sources of supply of natural gas. Nuclear explosives can accomplish this. And we admit there will be some radioactivity and this will cause some additional deaths, but think of the hazard to society of not having a supply of natural gas. Might not the result be even more deaths and a decrease in the quality of life?"

Let's not leave this decision to 'big brother'

This is really not an unreasonable set of questions to raise. Certainly society may, one day, have to choose the lesser of evils in a variety of situations. But, who, we must ask, should make such a vital decision for society—a decision which affects the quality of life not only in this generation but through genetic change, in all future generations as well? The Austral Oil Company? El Paso Natural Gas Corporation? The Lawrence Radiation Laboratory? The Bureaucrats of AEC headquarters? We say, emphatically, "NO." Enough of big brothers deciding what our fate should be. Big brothers in industry *and* government have already brought us to the sorry environmental plight we're already in—should we have much confidence in their erudite wisdom to solve the problems for us, behind our backs, in smoke-filled rooms of "experts"?

Such decisions, involving accepting a grave risk for some unknown persons in society versus a benefit ostensibly received by some, or all members of society, represent issues of the highest importance to the public. It should be their decision—arrived at after the fullest disclosure of all of the facts relevant to all aspects of the issue, the fullest participation of citizens, scientific or other, in such evaluation, and finally by referendum vote. Even this leaves much to be desired in view of the fact that a decision is being made to contaminate the planet irrevocably, to pollute the genetic inheritance of man for generations to come. But at least this is far superior to big brother making the decision, which he so much prefers to do, unhampered by the chill wind of informed public opinion.

The AEC and its flying road show

Do we see any evidence of a reasonable attitude on the part of AEC, or government, with respect to so fundamental an issue in a democratic society? None whatever. Instead we have overt evidence that the AEC is going to ram the radioactive natural gas program down the throats of the public no matter what they think. This is not our opinion; the record speaks all too clearly for itself.

In Colorado, AEC together with the Austral Oil, a prospective commercial beneficiary of the natural gas stimulation, launched so-called Project Rulison—to determine how much natural gas stimulation they could achieve by underground nuclear blasts. The citizens of the Colorado area involved had all kinds of reservations, questions, objections. Did the AEC fulfill its mission of being responsive to the public health and welfare in that instance? The AEC spends millions of dollars of taxpayer funds in so-called "education about the atom." They fly speakers and exhibits everywhere; they indoctrinate school children with exhibits and movies about the wonderful "peaceful" atom. But when the citizens of Colorado feel they may be victimized by an ill-considered AEC project in their very own community, will the AEC sponsor full public debate on all sides of the issues to demonstrate sincerity? Considering the public relations dollars the AEC is freely wasting on propaganda to sing its own praises, surely the cost of an honest, rather than one-sided, airing of the issues before Colorado citizens was indicated. No chance!

Colorado citizens feel AEC power

And when Colorado citizens finally brought suit to prevent flaring of the radioactive natural gas, did the AEC show itself as the impartial protector of the public health and safety? The most *primitive* elements of compliance with the spirit and letter of the Atomic Energy Act would certainly have required the AEC to assist in every way in the fullest, most open airing of the serious health issues raised by highly responsible citizens, scientists and others, in the State of Colorado concerning the Rulison gas-stimulation project. Did the AEC show such a serious understanding of its obligations under the Atomic Energy Act? They showed precisely the opposite.

Arrayed against the meager legal resources of the citizens concerned with their lives and property and health, the AEC brought in the high-powered legal staff, with all the resources of the entire U.S. Solicitor General's office, all at taxpayer expense. It is difficult to imagine that what actually was observed was the picture of AEC, a multi-billion dollar super-agency, using the tax-dollars of Colorado citizens in an effort to stifle their legitimate rights to life, liberty, and the pursuit of happiness. And in a callous display of contempt for the Colorado citizens, AEC paraded in a bevy of witnesses, all of whom presented a totally irresponsible one-sided picture of the radiation hazards question. And this at a time when the controversy concerning hazards of radiation is at its peak. Who paid, out of meager resources, to bring a few witnesses to present the other side of the hazards question?—the citizens of Colorado. Arrogance? Could more arrogance in the abuse of power be imagined?

This is no isolated instance of the irresponsible manner in which the AEC and its lieutenants treat the serious matter of public health hazards of radiation. In 1963 we agreed to investigate comprehensively the health impact of such AEC programs as the Plowshare Program. We were assured that an objective evaluation was desired. It would have been repugnant to us in the extreme to undertake that task at all unless we could evaluate objectively, and express our opinion about the Plowshare Program. We have worked hard on this, and other tasks.

Plowshare should be abolished

We conclude that the Plowshare Program is not safe for humans nor for the future of the earth. And we say so in no uncertain terms, with plenty of evidence to back our conclusions, that this is precisely what we mean about Plowshare. This program would be best abolished in society's interest. But we do not deny anyone else the right to hold and to express contrary opinion. For us to do that would be reprehensible. Is the Atomic Energy Commission equally reasonable? Fair? Let the record speak on this issue.

Dr. John R. Totter is the director of the Division of Biology and Medicine of the AEC. In this position he more than any single man in AEC is responsible for the fullest development and public exposure of information relating to the public health hazard of radiation and of AEC programs which can result in exposure of the public to radiation. It is unthinkable on such difficult matters that Dr. Totter would not expect controversies and controversial opinion to arise. But how does Dr. Totter react to any statement of evidence that radiation may be demonstrated to be harmful to man? We have some excellent examples.

In 1969 Professor Ernest Sternglass of the University of Pittsburgh published a paper suggesting that radioactive fallout from weapons testing during the 1950s might have caused as many as 400,000 infant and fetal deaths.[28] His manuscript was circulated to all of the Atomic Energy-supported bio-medical laboratories. Tamplin analyzed Dr. Sternglass' paper and felt that it was not likely that 400,000 deaths could be attributed to fallout; rather it appeared that poverty in the United States was a more likely explanation. However, reasoned Tamplin, Dr. Sternglass has served a very useful social function in raising the question of the real cost of radioactivity spread around the earth from weapons testing. He made the calculations himself and concluded that 4,000 infant and fetal deaths were more likely due to fallout than 400,000. And he prepared a paper for the *Bulletin of Atomic Scientists* presenting his estimate of 4,000 deaths versus Sternglass' estimate of 400,000.[29] Surely, Tamplin thought, it made no sense to criticize Dr. Sternglass' estimate without presenting a counter-estimate of his own.

An unusual suggestion from Prof. Teller

It so happens that at about this time the ABM debate was raging in the U.S. Senate, and ABM supporters were very fearful that Professor Sternglass' estimate of 400,000 infant deaths from fallout might be enough to sway a few crucial Senate votes. When Tamplin estimated only 4,000 deaths, Professor Teller and the Directorate at the Lawrence laboratory were overjoyed. The Tamplin manuscript was forwarded to the Joint Committee on Atomic Energy so that it could be distributed to worried U.S. senators. But Washington AEC headquarters did not share the joy of the Lawrence laboratory directors. What was apparently worrying AEC headquarters was a calculation that *any* infant deaths were being blamed on radioactive fallout.

Dr. John Totter wrote Tamplin a warm letter of congratulations for his contribution to this important problem. However, Dr. Totter suggested in his letter that Tamplin publish his criticism of Sternglass in one journal, and *separately* publish his own much lower estimate elsewhere. We have never figured out what thought processes could have possibly made such a suggestion reasonable, other than the possibility that even 4,000 deaths disturbed the AEC home office greatly. Tamplin declined to comply with Dr. Totter's suggestion.

Shortly thereafter Dr. Spofford G. English, AEC assistant general manager in Washington, called Dr. Michael M. May, director of Lawrence Laboratory, to ask how it was that Lawrence laboratory didn't have better "administrative surveillance" over Tamplin's publication activities. Since Gofman then still had the associate directorship for Biology and Medicine, the matter was referred to him for resolution.

Gofman was completely pleased about Tamplin's research activities and, above all, wasn't going to brook any interference with Tamplin's rights as a senior, independent scientist to do his work and publish it as he saw fit. A conference call was made to Washington, and Gofman and Tamplin tried to learn from Dr. Totter and Dr. English what was really bothering them. The result was an impasse. Dr. Totter and Dr. English simply didn't understand why Tamplin wanted to publish his own estimates along with his refutation of Sternglass' estimates. Gofman and Tamplin could say little else to the AEC officials other than if it were a whitewash of radiation hazards they were seeking, they had better seek elsewhere. So ended that particular episode—a highly informative one.

More recently Dr. Totter stated publicly, and is quoted in *Nucleonics Week* as follows: "Later, John Totter, head of AEC's Division of Biology and Medicine, said John Gofman 'was hired to make sure Plowshare could operate in a safe manner.' He is now attacking Plowshare and I see no reason that Lawrence Radiation Laboratory should want to continue his services."[30]

Dismissal for reaching the conclusion that Plowshare is unsafe and unwise? Hired to make Plowshare operate safely? To borrow from ex-King Farouk of Egypt, we can say the combined efforts of the king of hearts, spades, diamonds, and clubs couldn't make Plowshare safe! Why should Gofman be expected to accomplish such a miracle? Other than cancelling Plowshare, there is *no* safe answer for public health.

There is no description of this fiasco other than to say we have come a long way on the road to totalitarianism in science and technology. A la AEC the appropriate reward for reaching a scientific and personal conclusion at variance with their hoped-for preconceived conclusion is economic strangulation. Lest the AEC hide behind such outworn clichés as "Scientists should present only their scientific data, and not their personal opinions," we have a great deal to say on *this* issue. In our work on public health implications of atomic energy programs, we do our scientific work carefully, objectively, and sincerely. And out of such work we expect, and are assuredly expected by the public, to develop a personal opinion of the public health aspects of such a program as Plowshare. Formulation of such an opinion, based upon the best analysis of information at our disposal, is *the expected* culmination of our work. To leave that culmination step out would be irresponsible of us. And this is doubly needed in view of the immense, well-financed propaganda barrage emanating out of both AEC and Lawrence Radiation Laboratory concerning the "wonderful" accomplishments and benefits Plowshare can offer humanity. Is that not an opinion? Hucksterism, sales promotion—all based upon AEC opinion that something is good for society.

Conflicting opinions should be aired

We owe it to the public to do our work responsibly and to express our public health concerns openly, clearly, and audibly. We don't insist on dismissing the Atomic Energy Commission if they hold a public health opinion different from ours. *Both* opinions should go into the public forum for wide discussion and evaluation—that is, if we intend to do more than pay lip service to democratic institutions and procedures. Imagine our fate if every scientist has to feel that sanitized, cosmetized party-line opinions are required on matters affecting the public health if he is to survive with a place to work and earn his livelihood.

We can stand the heat of a debate on the merits of our public health opinion. If the AEC cannot tolerate contrary opinions, it would be far more appropriate to change the AEC, not to dismiss responsible scientists. We don't believe the AEC can stand facing the public health issues of radiation hazard in a really *open* forum. And we have good evidence for that belief.

How scientific findings are presented

When we reached our conclusions concerning the serious magnitude of the cancer and leukemia risk of radiation, it became appropriate, in the best tradition of science, to present the findings openly. In science, there is a propriety concerning how one announces new scientific findings. Either they are submitted to a scientific journal (which can require three months to a year before they are published), or they can be presented before a scientific society, which is the more usual scientific methodology. Following such presentation, the findings are published as proceedings or transactions of that particular scientific society. Every scientist in the world knows that this is appropriate and in the best scientific tradition.

Indeed, we had an unusual opportunity to present our findings at an especially appropriate scientific meeting, the national meeting of the highly respected body known as the Institute for Electrical and Electronic Engineers. The theme of their annual meeting in San Francisco in October, 1969, was "Nuclear Science and the Environment." If a more appropriate forum for presentation of our findings could be envisioned, we would appreciate knowing about it. And, further, we were not seeking an opportunity. The program committee of the Institute had sent us a special invitation to present a guest lectureship at this very important scientific gathering. And in the best professional and scientific tradition, we delivered our scientific presentation to that gathering.

As is customary worldwide, newspaper reporters attend such meetings, encouraged by the scientific and professional societies in the hope and desire that the public be accurately informed about new scientific developments. This is vital if the public is to be expected to understand why scientific work should be supported.

And were we in *any* sense inflammatory or derogatory in that presentation? Let us look at what we said concerning AEC in that presentation.

"We feel certain that the Atomic Energy Commission, the scientific and engineering community, and the electrical power industry are as concerned as we are to keep the environment safe for human habitation and to bring society the earliest possible benefits of the peaceful atom. And because we are certain of this, we urge all of these groups to join us in seeking an early revision downward by at least a factor of ten in the Federal Radiation Council Guideline for allowable exposure to the population-at-large."[31]

Now, we may be criticized for having believed that the AEC would behave honorably and in the public interest. But we were not prepared to believe that the AEC could react so violently and viciously in response to having the scientific and public communities learn the truth about the much worse cancer and leukemia hazards from radiation than anyone had previously realized—some 10 to 20 times worse than even some pessimistic prior estimates. But the evidence was there, and we had to present it. Surely the AEC realizes, we thought, that learning the truth is the only acceptable approach to such matters. They cannot seriously believe sweeping the truth under the rug can forever keep it from being known. The work we had done to arrive at our cancer and leukemia estimates from radiation exposure would obviously be done by someone else in the very near future, even if our evidence could have been suppressed by the AEC. They should have realized that.

A blistering attack from the AEC

But the promoter realizes very little, when he thinks his parochial interests are possibly threatened by the truth. And, so instead of joining us, considering the overwhelming evidence we had just presented at the respected scientific meeting, in the effort to achieve safe radiation standards, the AEC attacked us in a blistering unparalleled fashion. In so doing, the AEC has set back the prospects for a rational, safe development of atomic energy in this country more than they could have in any other manner.

The memory of the techniques of Adolf Hitler in coping with inconvenient truth are still fresh in the minds of most people who lived through that dark Nazi period. Tell a big lie, and tell it again and again as widely as possible. And Hitler knew the method worked, no matter that it be diabolical. Today's environmental rapists, bent upon their shortsighted view of gain, have not failed to appreciate the "useful" lesson from Hitler's "School for Scoundrels."

Thus, far from helping us protect the public health and welfare—a task the AEC itself had assigned to us—the AEC unleashed a blistering attack upon us, with slander, ridicule, denial —*with everything but any valid evidence refuting our findings*. Gone completely were the pious phrases of AEC about "We want you to tell the truth." Faced with a threat to its bureaucratic, parochial interests in selling its ware, the AEC clearly demonstrated that when the chips are down on questions of protecting human beings and their environment, the promotional, huckster role wins out handily over the public protector role.

We are not critical of the AEC, nor of the apoplectic reaction of its officials, sputtering and fuming insults at us for telling the truth. We were shocked, but gradually came to realize that they (the AEC officials) are victims of having been placed in a hopeless quandary by the Atomic Energy Act which assigned them two conflicting, irreconcilable roles—promoter and protector.

The proper approach is not to criticize the AEC unduly, but rather to take away from them all authority and responsibility for public health protection and all aspects of regulation of atomic energy. With their well-developed Madison Avenue type advertising and public relations enterprises, the AEC will be able to function as a sales organization for their diverse products and programs. But the public would then not have to fear that the salesmen will also be the guardians of the public health. We, as a society, know only too well the dire results of that particular combination in many, many areas besides atomic energy.

Some of the specific charges levelled at us by the AEC and hangers-on must be studied carefully, for by such study those who wish to *try,* at least, to preserve a livable world for themselves and their children can learn what they have in store for themselves from polluters of the environment:

"Unprofessional conduct—presenting their findings in the newspaper instead of as a scientist should," we were charged.

Well, as related above, we were guest speakers at one of the most eminent scientific meetings on the subject of "Nuclear Science and The Environment." Could a more professional and appropriate forum of presentation be imagined?

"Nothing new," it was said.

If there was "nothing new," why were all the AEC cannons roaring, even though ineptly aimed?

Or are we to assume that a radiation hazard of ten to twenty times as much cancer and leukemia as had previously been expected is "nothing new"?

What, we wonder, would be "something new" for the AEC officials? 100 times as much cancer? 1,000 times as much cancer? 10,000 times as much cancer? Just what would it take to worry the AEC about hazards to human beings?

"The experts considered all this already."

Indeed, unknown to us, a group known as a Task Force of the International Commission on Radiological Protection had reviewed the new evidence. When we saw their publication two months after our presentation, we were most gratified to find our numbers comparable to those of the Task Force.[32] Had we not done our work, they would have reached essentially our conclusions. So, the "experts" agree with *us.*

"Scientifically indefensible."

We are scientists. We know that the appropriate way to determine the validity of scientific conclusions is to present them so other scientists can review the findings and challenge them. We were delighted to make our conclusions available for other scientists to agree with or refute.

We are *still* waiting for the first bit of scientific evidence to refute *any* of our scientific presentation. The words "scientifically indefensible" shouted by wounded atomic hucksters can hardly be called scientific refutation.

"The benefits outweigh the risks—therefore, Gofman and Tamplin are wrong."

That's a real doll. The issue at stake is the risk of cancer and leukemia from radiation. Isn't it essential for AEC to at least *try* to calculate risk themselves, and *try* to estimate benefit, before announcing one outweighs the other? Besides, this has absolutely nothing to do with whether our risk calculations are correct.

"AEC Programs will never expose people to the allowable amount of radiation. Therefore, Gofman and Tamplin's calculations are wrong."

The first step is to reduce allowable radiation dosage

Our calculations describe how many extra cancers and leukemias will occur if everyone gets the allowable dose of radiation. And the first thing we said in our presentation was that we should take steps immediately to see that they never do. And the one *sound* step to take is to reduce immediately the allowable amount before such exposure becomes possible.

Indeed, it baffles us to hear AEC and nuclear proponents outdoing each other with: "We won't give people 1/10 of the allowable dose; we won't give people 1/100 of the allowable dose; we won't give people 1/10,000 of the allowable dose."

"Marvelouser and marvelouser," we outside the AEC "Alice in Wonderland" world would say. "Since you don't *need* to give people the radiation from AEC programs, then our proposal for safe standards is in no way a thwart to your super-safe programs. Join us in achieving Federal codification of safe standards for radiation standards."

The resulting "fume, sputter, splat" gracing the air at AEC headquarters in Germantown, Maryland makes it quite clear that the AEC officials know very well that reducing the allowable radiation dosage will indeed interfere with their pet promotions.

Look back again at Professor Teller's remarks in this chapter.

Faced with a total inability to present any scientific evidence to counter our leukemia and cancer estimates from radiation exposure, the final empty trick was attempted by AEC and its sycophants. Attack not the scientific findings, but attempt to destroy the character and propriety of those who make the scientific findings which disturb the promoter. Thus we soon heard we had presented our findings to the wrong scientific meeting. Why wrong? Because these were engineers who really were in no position to determine whether the results were believable. This allegation was dumbfounding to us. The top electronic and electrical engineers decide to devote their annual scientific meeting to a serious discussion of the environmental impact of the nuclear science they conduct, and we are told these engineers are too stupid to be presented with evidence for their critical evaluation. We shall leave the engineers to muse over this.

Congressman Craig Hosmer of California, presented the same "criticism" in a conference. Why, he asked, didn't we present the findings to our scientific peers—for open review and criticism? Indeed, why not? We had just done so, before the electronic engineers, and furthermore in our testimony before the Muskie Committee we had stated:

> Today we have presented your Committee with much evidence indicating that current radiation exposure guidelines are indeed dangerous—much too high. It would be naive of us to believe that our recommendations will be received with enthusiasm in all quarters. To the best of our ability we have endeavored to present the truth. Our calculations, our evidence, may, upon critical examination by others, prove wrong in minor respects. We doubt they will prove wrong in any major respect. The sharp cutting edge of scientific criticism, with all the evidence placed squarely in the open forum, will demonstrate any fallacies, will show where additional evidence is needed, and where errors have been made.[33]

Critical examination by our peers was *precisely* what we wanted, expected, pleaded for—what was this nonsense about "no peers"? Here we had presented a body of evidence where every scientist in the world could review, criticize, recalculate, or do anything he wanted, to agree or disagree with us. And who would prevent that scientist from making a public presentation of his findings? We couldn't see anything to stop criticism of our findings. We suspected, and are now certain, that the only reason for the "peer" story going on for months was simply a device to try to discredit damaging evidence without any counter-evidence available—one of the oldest dodges in the world, and so revealing.

It seemed worthwhile to find out if the Atomic Energy Commission really had any concern about where or to whom we had presented our findings, or whether they were simply downright unhappy that true facts concerning radiation hazards were now out in the open. We decided to test the "presentation to peers" concept once and for all. So at the Joint Committee on Atomic Energy Hearings, we asked for a complete hearing before truly capable peers—the best scientists in the land. We said we were willing to present our case to a jury of such peers. (See our challenge in Chapter 2.)

And what did Congressman Hosmer say in answer to our offer to have the *best* scientists, impartial and with no axe to grind, judge us and our findings concerning the serious hazard of Federally allowable radiation doses?

Congressman Hosmer answered, "That sounds like the same method used to authenticate the Dead Sea Scrolls."

Indeed, Congressman Hosmer, *you* wanted peers—and when peers are offered, you're off on the Dead Sea Scrolls. How so?

And the AEC, have they been heard from? Are they willing to debate before an impartial jury? No word. No word at all, other than a call for our heads! We rather doubt this debate will ever be held, and if not, it won't be because of *our* reluctance!

6 Tragedy on the Colorado Plateau

"Atomic Energy is the Safest Industry in the Land" is the theme song of the atomic energy promoters.

We are constantly bombarded with the twin platitudes, "We understand radioactivity hazard better than any other" and "No industry has a safety record like the atomic energy industry." Beautiful, reassuring. But is it true? To understand the atomic energy enterprise and what it means to people, we dare not start in the middle, for by then so very much has been well tidied up, carefully swept under the rug by a devoted army of public relations "educators." No, it is necessary to start at the beginning. And the beginning of this industry's story starts with one substance—uranium. Without uranium, there is no atomic energy enterprise. Uranium must be won from the earth, by old-fashioned mining, carried out wherever sufficiently rich deposits are found. Once one understands the uranium mining story, one realizes that conditions are not really much different from the rugged pioneer days of the yellow metal, and the value of a human life not appreciably greater.

Rich uranium deposits occur in several areas of the world—the Congo, formerly ruled by Belgium, being one of the significant areas with rich deposits. In the United States the Colorado Plateau provides a reasonably abundant supply of uranium ores. Since nuclear bombs are today's military sine qua non, it is understandable that the U.S. Government would be loathe to depend upon extra-continental supplies of this critical element for the nuclear era. So uranium mining had to be developed on the Colorado Plateau. A cheap supply of uranium is the name of the game, especially when, as we shall see later, a determined effort is in progress to make uranium-based nuclear electricity *appear* economically attractive. Cheap uranium starts with lack of concern for the fate of the men who mine uranium ore.

Uranium miners on the Colorado Plateau have been dying of lung cancer in what can best be described as an *epidemic* of this dread disease. Is the reason known? Yes, they have had their respiratory tracts and lungs exposed to a powerful cancer-producing stimulus—the alpha particle radiation coming from "radon-daughters." What are radon-daughters? Uranium, as it occurs in natural deposits, decays slowly through steps to radium, radium in turn to radon. And radon, which disappears quickly (about half of it disappears in three days), gives rise to a sequence of radioactive by-products known collectively as *radon-daughters*. While the long persisting radium is present, there will be radon and radon-daughters. Thus, anywhere uranium is mined, radon-daughter exposure is an expected hazard. The degree of hazard depends upon how well the mine is ventilated to sweep out radon, a gaseous element. For without radon there are no radon-daughters, and, hence, no hazard.

Wagoner* and his colleagues of the Occupational Health Program of the U.S. Public Health Service have established beyond a shadow of medical doubt that the epidemic occurrence of lung cancer in uranium miners on the Colorado Plateau is undoubtedly due to exposure to radon-daughters.[34] And thus the uranium miners, possibly an ultimate 500 or more of them, represent a major group of casualties of that "safest" of all industries, the Atomic Energy industry—via fatal lung cancer. And this is only the beginning of the story as we shall presently see.

One might feel sad that such an "unexpected" result of the development of atomic energy has occurred, that it is unfortunate, and say, "How was anyone to know this would happen?"

*Joseph K. Wagoner is Statistician, Epidemiology Branch of the National Cancer Institute.

But the truth of the matter is quite otherwise. From knowledge available to everyone concerned it was, from the start of the rush to mine uranium in the 1940s, known that with exposure to radon-daughters, there would be occurrences of lung cancer among the exposed miners. There was no need to discover this medical fact on the Colorado Plateau after 1940. For medical texts recorded this result—lung cancer from uranium mining had occurred decades before in uranium miners in Schneeburg, Germany, and Joachimsthal, Czechoslovakia. And for decades, every medical authority realized that radon and its radon-daughters represented the culprit. But we had to repeat the unnecessary tragedy of miner's lung cancer on the Colorado Plateau, and we are likely to continue to do so unless stringent steps are taken to avoid it.

There is a commonly-agreed measure of exposure to radon-daughters called "the working level" of radon and radon-daughters. It specifies a certain concentration of the alpha-emitting radionuclides per volume of air. The numerical value of this unit need not concern us here, except to state that miner exposure can be described in terms of the "working level" in his mining environment and the number of months he has been working at such levels of radon-daughter exposure. One simply multiplies the working level times the number of months (8-hour days, 5-day weeks) to get what is called working-level-months, or W.L.M. units of exposure. Individual mines have operated under conditions differing widely from each other in the working-levels of radon and radon-daughters present. And thus some miners, in a few years of employment, accumulated a large dose in working-level-month units; others much less.

Lung cancer epidemic could have been prevented

The Wagoner report appeared in 1965, indicating radon-daughter exposure as the cause of the observed epidemic of lung cancer among the miners. A ripple of dismay spread through the atomic energy community. And that dismay grew to a roar as it became widely known that such an epidemic could have been

Tragedy on the Colorado Plateau 131

fully anticipated based upon knowledge long available. As is usual in such matters, the first task is buck passing—for who among governmental officials or agencies will accept the blame for a rash of fatalities resulting from benign neglect by "someone"? The Atomic Energy Commission maintained that uranium mining per se, and the conditions therein, were not within its province of authority. Others disagreed.

But the whole episode became too difficult to sweep under the rug. Thus, the Joint Committee on Atomic Energy decided that a full set of open hearings on this tragedy must be held in order to understand what had happened, why it happened, and how it might be avoided in the future.

For those miners dead of lung cancer, no ameliorative measures are available. For hundreds of miners who have already been exposed to the cancer-producing alpha rays, there also is no ameliorative measure —they need only to wait several years to develop lung cancer. Nothing will save them from their fate.

The set of hearings held by the Joint Committee on Atomic Energy are entitled "Radiation Exposure of Uranium Miners."[35] In two volumes they represent a remarkable account—evidence of how utterly irresponsibly human beings can behave where the health and welfare of humans impinges upon the "economic" health of industrial enterprise. And what has happened since those hearings is even less comprehensible or defensible! Those hearings contain a most interesting document entitled "Guidance for the Control of Radiation Hazards in Uranium Mining" issued as Special Report No. 8 of the Federal Radiation Council. We shall return to this later.

The hearings also contain testimony presented by one witness after another relating to the expectation that lung cancer must occur with radon-daughter exposure, and suggesting that the way to prevent additional new, unnecessary deaths would be to ventilate the uranium mines in order to achieve a drastic reduction in radon concentration in mine air, and hence a drastic reduction in lung exposure to radon-daughters.

One wonders why it took two volumes of hearings to arrive at some obvious conclusions, until one realizes that dollars are involved. Cleansing the air of uranium mines costs money, and if money is spent to save human lives, the profits from uranium mining are lessened. But there were even deeper issues at stake. Uranium is the requisite element as fuel for nuclear reactors. And nuclear reactors were being touted as the future salvation with respect to meeting the electric power needs of the nation. Nuclear reactors depended, in large measure, for acceptance upon the demonstration that economically they could compete favorably with coal, oil, or gas, the major fossil fuels used to produce steam to drive electrical generators. Obviously, the more costly uranium fuel, the less attractive nuclear power would appear compared with fossil fuel power. And this to atomic energy promoters, such as AEC and the Joint Committee on Atomic Energy, would indeed represent a tragedy—for them. And once one understands this primary fact, it becomes ever so much easier to understand why two volumes of hearings were required. The really pertinent evidence could have been presented in a third of one volume. The gobbledygook of obscuring the issues occupied the remaining one and two-thirds volumes.

Obscurantism replaces a simple solution

Simple problems, with obvious answers, ought, by reasonable men, to be solved with dispatch. Radon-daughters represent a known cancer-producing poison; the miners were being exposed to this poison. The miners were dying of an epidemic occurrence of lung cancer. The answer: Clean up the air of the uranium mines, and then the miners won't die of lung cancer. Very elementary. But why seek elementary solution when by jargon and ridiculous rhetoric, more complicated solutions can be made available? And true to expectation in so many aspects of atomic energy development, the jargon and confusing obscurantism prevailed—long enough to fill two volumes of hearings in the august halls of the Joint Committee on Atomic Energy. The relevant testimony was short and to-the-point. Only it appears no one was listening to that.

Thus, Chairman Holifield explored in great detail how one might be sure of precisely the radiation dose received by a particular miner. After all, he reasoned, knowing what's in the air the man breathes is only indirect evidence of what really was the exposure at the lung surface to alpha rays. And then he wondered in those hearings how accurately was the concentration of radon in the air of each mine known? How could we be sure a particular man's cancer was really due to radon exposure if we didn't know his *exact* dose of radiation to the lungs? Technically, every question is correct. From a public health protection viewpoint, they represent trivial, irrelevant, nit-picking in the extreme.

What was abundantly clear to everyone was that lung cancer was indeed occurring in relation to radon-daughter exposure. And further, it was clear that those miners who worked in the dirtiest mines and thereby accumulated the highest working-level-month exposure in the course of their mining were the ones with the highest rate of development of lung cancer. Reasonable public health practice would have reached the conclusion that at lesser exposure, there would be fewer lung cancers in direct proportion to the lower exposure to radon-daughters. But this would be too reasonable. What, it appeared, was desired by the promoters was direct *proof* of occurrence of extra lung cancers at each working-level of exposure, before condemning such exposure as unsafe.

The great health physicist, Dr. Walter S. Snyder* pleaded at the hearings as follows:

> Those who prefer to base radiation protection on a threshold hypothesis, which is just as unproven and just as uncertain and unsupported by data as is the linear hypothesis, often charge that the linear hypothesis is too conservative. There is no evidence, as illustrated above, to show that it is conservative at all. However, one may wonder why it is considered so undesirable to use a conservative criterion where human life is in question. Surely if the linear hypothesis is conservative and is not in con-

---

*Dr. Walter S. Snyder is Associate Director, Health Physics Division, Oak Ridge National Laboratory.

flict with the data that are available, this is a point in its favor. When human life is in the balance, it would seem that conservatism in safe-guarding these lives has much to commend it.[36]

Our impression is that Dr. Snyder wasn't being listened to at all insofar as any impact could be discerned as a result of his eloquent testimony. We have been criticized for statements made by us in writing that atomic energy developers demand the corpses of victims before they will cease doing that which is unsafe. You, the prospective victims, must prove with corpses that we're hurting you; otherwise we may proceed to irradiate you without any requirement of our proving safety of our promotional atomic energy activities. Yes, we *have* made such statements concerning several aspects of atomic energy development. And nowhere is the statement more appropriate than in criticism of the handling of the uranium miner problem.

Men die as FRC pleads for more 'evidence'

Instead of accepting Dr. Snyder's reasonable public health conservatism to protect the lives of the miners, the Federal Radiation Council concluded that, at the highest level of exposure in working-level-months, lung cancer was clearly in excess—but more evidence was required to be certain that lung cancer was truly occurring in excess at moderate or low working-level-month exposure. More evidence required by whom? By the uranium miner who stood to die from moderate exposures? Or by the mine owners and operators who would bear the costs of ventilating the mines?

Indeed, we have recently analyzed the very data which were already available for the 1967 Hearings of the Joint Committee on Atomic Energy. Our conclusion is that quite adequate evidence was available *then* that excessive lung cancer was occurring not only at the high working-level exposure, but at intermediate and low working-level exposures as well. And more than that, if anything, the evidence indicated strongly that one unit of exposure was more effective in producing lung cancer in the low exposure range than in the high exposure range. So, even though, as Dr. Snyder pointed out, one should always err on the side of protecting the public where knowledge is scant, in this particular case even the knowledge then available was adequate to arrive at a conclusion that lung cancer could be expected at low or moderate radon-daughter exposures, and more lung cancer could be expected at high radon-daughter exposures. But unwelcome answers are rarely sought with diligence.

The false hope of a safe threshold

In a now predictable fashion, the bad pennies of atomic energy development have a way of showing up again and again. We refer to the ever-present bad-penny "hope" of atomic energy developers that, "Maybe a safe threshold exists; maybe below *some* level of radon-daughter exposure there won't be extra cases of lung cancers among the miners." We have pointed out elsewhere in this book the cruel hoax aspects of translating such unfounded hopes into policy-setting with respect to exposure limits of human beings to radioactivity. And we have pointed out that every responsible body, including the International Commission on Radiological Protection and the Federal Radiation Council, refuses to consider such safe thresholds in practice, simply because no convincing evidence has ever been provided to justify such hopes.

Why, then, didn't the Federal Radiation Council insist on similar conservatism for the uranium miners? No answer.

And let us consider some of the dialogue that went on at the famous Joint Committee Hearings.

Professor Robley Evans,* speaking to Congressman Hosmer, said: "I am perfectly glad to turn the statement around the other way. I believe, in a positive sense, that 1 to 3 working-levels and a total accumulation of 300 to 400 working-level-months is innocuous to man."[37]

The old "safe threshold" concept at its best! Reassurance that below 300 working-level-months no lung cancers would occur

*Dr. Robley D. Evans is Professor of Physics, Massachusetts Institute of Technology.

in the uranium miners! Apparently the two chief promoters of JCAE, Chairman Holifield and Representative Hosmer, were inordinately pleased with this totally unsupported, unjustified optimistic prediction. Why shouldn't they be? Such a prediction of zero risk meant that the heat was momentarily off with respect to cleaning up the air which the unfortunate uranium miners had to breathe.

But the enthusiastic reception of the hopeful, though negligent, prediction of no hazard was to be very short lived for one reason —facts have a way of becoming evident to everyone! Just two short years later, in one of the most important papers ever published, Drs. Lundin and Archer* and their colleagues presented, *not* hopeful glowing optimistic predictions, but the grim, horrible facts. Not only were lung cancers occurring at the 300 working-level-months predicted by Professor Evans to be "innocuous"—they were occurring at FOUR TIMES the expected rate even at 180 working-level-months.[38] Where safety had been predicted based upon unsupported blue-sky optimism, tragedy was the result, in the form of a massive increase in frequency of fatal lung cancers among the unsuspecting uranium miners— men who rely upon scientists to know enough to help protect them from unnecessary injury.

The AEC doesn't learn from mistakes

It must be obvious to everyone that no criticism is intended either of the hopeful congressmen or super-optimistic scientists without sound scientific evidence—that is, criticism of any willful neglect. Not at all. But who is to protect the public and workers in industry from arrogant confidence in fanciful hopes? Who has apologized to the uranium miners for the grievous injury, even though not intended? Whenever one predicts confidently that some procedure or action is safe, and that very action leads to

*Dr. Frank E. Lundin is associated with Occupational Studies Unit, Division of Environmental Health Science, National Institute of Health. Dr. Victor E. Archer is with Occupational Health Program, National Center for Urban and Industrial Health, Bureau of Disease Prevention and Environmental Control.

deaths, one can, as a minimum, hope that some humility is learned thereby, even if the human lives are irreversibly lost. Certainly the uranium miners and their families must, as a result of this grave blunder, have lost confidence in science, scientists, technology, and in governmental "watchdogs."

Is there evidence that the lesson of humility and shame over the tragic, unnecessary deaths of uranium miners was learned by the congressional AEC hawks? No. The *opposite* is the case, and this deserves careful scrutiny.

The Secretary of Labor, then Willard Wirtz, a very responsible man, with a high concern for the laborer's welfare, apparently studied the records of the uranium mining industry quite carefully, and apparently appreciated quite deeply the sound testimony presented in the uranium miner hearings. He must further have appreciated quite well the sound advice provided by Dr. Walter Snyder. So, whereas the Federal Radiation Council in its September 1967 report recommended allowing 1 working-level in the uranium mines—a level that would approximately *double* lung cancer in miners after eight years of work in the mines, the Secretary of Labor was not satisfied with such lax standards and issued a Labor Department set of Radiation Standards for Mining, requiring that by January 1, 1969, all levels in uranium mines be reduced to 0.3 working-level—30% of the recommendations of the Federal Radiation Council. The judgment and courage of Mr. Wirtz deserves the highest commendation and praise.

FRC's unique benefit vs. risk approach

Let us contrast the order of Mr. Wirtz with the factors that occupied the Federal Radiation Council. The FRC said that major findings of immediate interest to the derivation of guidance for radiation protection in uranium miners are as follows: (1) Uranium is currently the basic fuel needed for the development of nuclear energy, and all projections point to an increasingly important role for nuclear energy in meeting national electric power requirements; and (2) Uranium mining is an important economic asset to the states in which the ore is mined. In addition to the value of the ore, mining provides important opportunities for employment.

What profound human concerns are expressed in these two statements of the Federal Radiation Council! The country needs cheap uranium for electric power generation (so that nuclear power can compete *economically* with other methods of electric power generation). Obviously protection of the miners from lung cancer might increase the price of uranium, so death from lung cancer for miners helps nuclear electric power appear cheap. This is a very interesting FRC benefit versus risk approach.

Congressman Holifield attacks Mr. Wirtz

And the second FRC statement that uranium mining is an *economic* asset to the states in which ore is mined—what is the economic asset worth to a uranium miner dead of lung cancer, and to his family? Strange, these benefit-risk estimates.

Obviously, Mr. Wirtz was not especially impressed by the FRC considerations. He ordered more than a three-fold reduction in the standards found acceptable by the FRC.

But the battle to prevent the uranium miners from dying unnecessarily of lung cancer was by no means over. On October 30, 1969, Congressman Holifield, the champion Congressional protector of the AEC, vented his anger over the totally responsible action of Mr. Wirtz. We cannot refrain from quoting Congressman Holifield:

> Secretaries of the Departments of Health, Education and Welfare, and some others. And these people are laymen. They are not technically qualified, and this is one of the things that has bothered this committee because we say the incident last year on the standards for the operation of the (uranium) mines which was set, as this committee determined, on an emotional basis rather than on a scientific basis.
>
> It is no secret that I said to Secretary Wirtz at this time that this is exactly what he was doing.[39]

Well now, Congressman Holifield, is reducing the radon-daughter level *downward* from a value capable of doubling the lung cancer rate regarded by you as emotional? The action of a layman?

If Mr. Wirtz' action in trying to protect the lives of uranium miners from the dreaded lung cancer is emotional, we applaud his emotional action, and say that we need more such emotional men in government. And we might all look forward to the day when Congress has some more emotional men who worry more about saving the lives of uranium miners than keeping the price of uranium low enough to promote nuclear electricity at high *human* cost!

The tragic, totally unnecessary occurrence of an epidemic of fatal lung cancer in uranium miners of the Colorado Plateau is a blot upon the development of atomic energy in the United States. But this is only the *beginning* of the real tragedy.

When uranium is extracted from its ores, there is left millions of tons of so-called uranium mine tailings. Everyone involved in this industry knows that the radium present in the uranium ore is left behind in these tailings. And where radium goes (it lasts for thousands of years), there will be radon gas and the inevitable radon-daughters—those same radon-daughters which produced lung cancer in the miners.

Uranium mine tailings used to fill home sites

Once the uranium was extracted from such ores, the tailings, according to the Atomic Energy Commission, were no longer their responsibility. They defined their responsibility to end when no more than a certain amount of uranium was still present. Very convenient. Who did worry about these uranium mine tailings with their still-present poisonous radium content? Apparently no one worried very seriously, because some enterprising contractors were able to obtain these tailings as a cheap fill to be used in housing and building developments in Grand Junction, Colorado, and elsewhere. Homes were built upon these tailings. No one knows precisely how many homes, but estimates range in the neighborhood of 3000 or more.

Radon gas escapes from such tailings into the homes built

upon them, and then the inhabitants breathe the radon gas *plus* its deadly radon-daughters. And the women and children don't spend just the eight hours per day in their homes that the uranium miners do in the mines. They may spend 16-24 hours per day breathing the lung cancer-producing radon-daughters. Indeed, it is entirely possible that for a uranium miner and his family living in one of these homes constructed upon uranium mine tailings, the miner may be the safest member of the family. His family may get more radon-daughter exposure in the home than he does in the mine.

Homes now being condemned due to radioactivity

The precise magnitude of this fabulous blunder in mishandling of wastes of the atomic energy industry is only slowly being unfolded. A slow, leisurely program is in progress to determine which homes are built upon uranium mine tailings, and how high the radon and radon-daughter levels are in the individual structures. We already know some high levels have been discovered—high enough to cause the homes to be condemned from future human habitation.

Why is the identification of which homes represent a mortal hazard proceeding slowly? In part, because of a human frailty. Many people have their life savings invested in the homes built upon uranium mine tailings. If they admit to owning such a home, the possibility exists, they feel, that the property value will plummet, and their life savings will be gone. So they hope to keep silent about the fill underneath their homes, hoping to sell them before it is discovered that they have a radon-daughter hazard.

These people should not be scorned. They have been innocently victimized. The real question to ask is where is the sense of responsibility of the U.S. Government in this entire scandal! Uranium mining is at the source of the entire hazard. And the uranium mined was largely for the purpose originally of producing nuclear weapons, ostensibly for the defense of the people of the United States, and more recently for the development of nuclear reactors for electric power generation. Since these uses of uranium were presumably for the benefit of all Americans, why are a few thousand families bearing the inhuman cost of a stupid blunder of several segments of the entire atomic energy complex? It is not a question of placing the blame on anyone. Rather it is a question of the need for human compassion where neglect prevails.

The President of the United States should long ago have declared Colorado a disaster area, victimized inadvertently by a national atomic energy enterprise. And every home built on uranium mine tailings should have been measured for radon. Where radon is found, the homes should have been moved or destroyed and the inhabitants compensated by the Federal Government. Anything less is sheer inhuman neglect, a product of our times. Yet we see no disaster area declared, and the radon-daughter exposure continues.

Will radioactive natural gas be next?

Colorado, one of the gems of the beauty that is America, has indeed entered the atomic era with vigor, thanks to the Atomic Energy Commission activities plus the uranium mining industry. A few miles outside Denver, the Rocky Flats plant of AEC handles fantastic quantities of the lethal plutonium, capable of producing lung cancers at doses of a millionth of a gram. The Plowshare enthusiasts, totally casting aside the protest of Colorado citizens, have fired one of their underground nuclear bombs to "stimulate" natural gas production, which gas as related elsewhere will be radioactive and, hence, poisonous. The uranium miners are dying of lung cancer, and homes of unknown numbers of human beings may be more of a hazard than the uranium mines because of the deadly mine tailings upon which they are built.

Along with its pride in its skiing and world famous music festivals at Aspen, Colorado now sports bumper stickers on its cars which read, "Come to Colorado, the playground of the Atomic Energy Commission." Possibly, with a gas mask!

7 Nuclear reactors

Since we were at the Lawrence Radiation Laboratory, in Livermore, California, our major concern had been with nuclear explosive devices. The reports that we had written were related to the dosage to man that would result from the application of nuclear explosives in the various ill-conceived engineering projects of the Plowshare Program. At the same time, we knew that our general approach could be readily applied to the nuclear reactor industry.

We began to turn our attention to the burgeoning nuclear power industry subsequent to a symposium in March, 1969, that represented the dedication of our long-awaited Bio-Medical building at the laboratory. The title of the symposium was, impressively, "The Biological Implications of the Nuclear Age." As symposiums go, it was one of the worst. Professor Joseph Rotblat,* reviewing the published proceedings in the May 8, 1970, issue of *Science,* stated: "To sum up, the reviewer is somewhat doubtful about the value of publishing these proceedings." Perhaps the main significance of the symposium was that one of the few non-AEC people who attended was Dr. Dean Abrahamson** of the University of Minnesota.

Dr. Abrahamson and a small group of individuals in the Minnesota Committee for Environmental Information had helped

---
*Dr. Joseph Rotblat is Professor of Physics, London University at St. Bartholomew's Hospital Medical College.

**Dr. Dean E. Abrahamson is Associate Professor of Anatomy and Laboratory Medicine, University of Minnesota Medical School.

fight a successful battle to get the State of Minnesota to set its own standards for radioactive releases from the proposed Monticello power reactor of the Northern States Power Company (NSP). The State's standards were fifty-fold more restrictive than those of the AEC.

The AEC claims that Congress has pre-empted the right of the states to set such standards and forced NSP to take the State of Minnesota to court over the issue. In August 1969, at the Governors' Conference held in Denver, the governors voted unanimously to support the position of the State of Minnesota. Subsequently, a number of states have joined in the case with Minnesota and others have filed *amicus curii* ("friend of the court"). Finally, in June 1970, the State of Maryland proposed a set of standards that are a hundred-fold more restrictive than those of the AEC. As we shall see subsequently, these are wise decisions on the part of the states because the AEC standards are, to put it mildly, ridiculous.

Taking a closer look at the nuclear power industry

Because of our conversation with Dr. Abrahamson, we decided that we should take a closer look at the nuclear power industry. In August 1969 we received three invitations to attend conferences related to nuclear electrical power reactors. Dr. Abrahamson was instrumental in creating the invitations to a conference at the University of Vermont in September and to a conference at the University of Minnesota in October. Dr. Barry Commoner, of Washington University in St. Louis, was responsible for our invitation to a symposium at the annual meeting of the American Association for the Advancement of Science (AAAS) in December.

The conference in Vermont was sponsored by the governor, Deane Davis. It was a precedent-setter in that it was originally planned as an AEC propaganda sideshow to be staged by the AEC's redoubtable traveling circus. However, the Lake Champlain Society and the Conservation Society of Southern Vermont managed to get the governor to influence the AEC to set aside a portion of the program for a direct confrontation with its critics.

Tamplin attended this conference and was one of a panel of four who confronted an equal number from the AEC. We got our first real impression of the oppressive nature of the AEC juggernaut at the meeting. Because of the way science and engineering is funded in the country, most knowledgeable experts in this field are in the employ of the AEC or the power industry. It wasn't easy for the conservation groups to find their experts. They had to look for their people from among those who developed their expertise in their spare time or who were no longer with the AEC or the industry. It seems that few people will publicly criticize their employer. The AEC, on the other hand, was reported to have brought 36 people to Vermont, including three commissioners and even the chairman, Glenn Seaborg. Some were flown in by Air Force planes.

The AEC road show

Except for the panel discussion, this was a well-planned propaganda campaign paid for by the taxpayers. The AEC had a number of manned exhibits, like a carnival. It bused school children to talks and displays on the wonders of atomic-generated electricity. Moreover the panel of critics was put at a disadvantage because the AEC show began a day earlier and then the panel discussion followed a series of talks by the AEC "experts." The deck was really stacked against the critics. Seeing the taxpayers' dollars used against the public in this way, one quickly comes to recognize that the best way to fight city hall is to vote its occupants out of office.

Nevertheless, even with everything set in its favor, the AEC lost in this confrontation. It lost because the people of Vermont could see through the shallow presentation and recognize the evasiveness.

Tamplin went to Vermont with the sole purpose of trying to get the AEC to present an estimate of the biological effects of exposure at the FRC radiation protection guideline. The imminent failure of this attempt became obvious during the prepared remarks of U.S. Atomic Energy Commissioner Theos Thompson, who stated his commission's incredible position:

> The maximum permissible levels are established at a point such that there is no accepted evidence of any sort of genetic or tissue damage to any human being exposed to these maximum levels.[40]

Contrast this statement with that of Dr. E. Eric Pochin,* Chairman of the International Commission on Radiological Protection (1962-1969):

> The position regarding occupational exposure, and that of members of the public, has clearly become very much more complex with increasing evidence as to the possibility of occasional harmful effects even at low doses or dose-rates, so that it is no longer a question of recommending the levels of exposure which are safe, in any absolute sense, but those which can be considered as appropriately safe for the circumstances in which they need to be received.[41]

During the subsequent panel discussion, the following occurred:

> GOV. DEANE C. DAVIS. The next question is addressed first to Dr. Storer** and then to Dr. Tamplin. The question is: If human population were subjected to the maximum radiation exposure allowed under the radiation protection guide, would this result in an increase in cancer and a reduction in lifespan?
>
> DR. JOHN B. STORER. This might happen in theory, but in practice, it would not be detectable. That is why the standards were set at that level. With respect to the effect on lifespan—the late changes that occur that lead to shortening of the lifespan in animals develop after a long latent period. The lower the dose, the longer the latent period. If the dose is so low that the latent period is longer than the life expectancy, then the effect never really has an opportunity to manifest itself. So, while theoreti-

---

*Dr. E. Eric Pochin is Director, Medical Research Council Department of Clinical Research, University College Hospital Medical School, London.

**Dr. John B. Storer is associated with Bio-Medical Research, Oak Ridge National Laboratory. He was formerly Associate Director, Division of Biology and Medicine, AEC.

cally, life shortening could occur, but in either case, the changes would not be detectable.

DR. ARTHUR R. TAMPLIN. I am just curious as to what one means that they would not be detectable. They would be present, but you would not be able to detect them.

DR. STORER. Not with the population size available and with the inability to control other variables—there is simply too much noise in the system to detect this particular effect.

DR. TAMPLIN. Well, then you must have some idea of what that effect might be numerically.

DR. STORER. Sure, you arrive at this by back extrapolating from studies done at high doses. You put up a conservative model, and you assume for this case that it scales linearly. Then you extrapolate back and get into the numbers game, and you can say theoretically that there will be five more cases or ten more cases, but, in view of all the errors in the system, you will never be able to detect this effect statistically.

DR. TAMPLIN. The question was not with your ability to detect it, if indeed it cannot be detected, it is what is the numerical value, theoretically. Obviously, if you cannot detect it, there is no other way you can arrive at it. If the present levels of radiation protection guidelines have been set, as Dr. Russell* said, by a group of competent scientific individuals who have weighed this situation carefully, then that must mean that they have an idea of what the precise effect would be, theoretically at least, on a scientific basis.[42]

The AEC spokesmen refused to answer this question at Vermont, or anywhere else for that matter. The message which these spokesmen chose to convey at Vermont was that exposure at the allowable dosage was safe. Then they indicated that nuclear reactors would expose individuals to only a small fraction of this allowable dosage. Thus, they tried to create the impression that the reactors were more than safe. We knew that there was no valid basis for indicating that the allowable dosage was safe, and we had only their word for what would be the resultant dosage from reactors.

Upon Tamplin's return from Vermont, we came to two con-

*Dr. William L. Russell is in the Biology Division, Oak Ridge National Laboratory.

clusions: (1) The AEC had been so shallow and glib that it must have been avoiding some serious problems with respect to the nuclear power industry; and (2) if we were going to bring these problems to light, we were in for a dirty fight. Both impressions were correct.

It was one month later, October 1969, when Tamplin went to the symposium at the University of Minnesota. It was only at the insistence of Dr. Abrahamson that we were invited to this symposium. Prior to going, Tamplin received the following admonition from the symposium chairman, Dr. Harry Foreman:*

> I am very much looking forward to hearing about your work but I do have some concern as to how you handle your data in relationship to the ICRP standards. The atmosphere in Minnesota is highly charged vis-a-vis nuclear energy and doubts voiced by reputable scientists may well result in a furor that could not only drive nuclear power plants from the state forever, but also any radioactive material used for research and diagnosis. You should be aware that a serious challenge to the safety of the ICRP standards is highly consequential here and should not be made lightly. Of course, if you believe from the basis of your data that the public would be harmed from discharges at the maximum permissible level, then you are free to indicate this at the sessions. You might want to talk this over with Dean or any of us before the sessions.[43]

It seems that there was a strong desire to have a meeting that would discuss only the wonders of nuclear reactors. This desire seems to have extended over into editing the proceedings of the symposium. A striking example is discussed in the subsequent chapter on waste disposal where Dr. Foreman refused to let Dr. M. King Hubbert insert some very important material into his paper. At the same time, individuals were permitted to alter their statements in the discussion section and were even allowed to withdraw their comments.

If you recall, we stated earlier that the State of Minnesota had proposed a set of regulations governing the release of radionu-

---

*Dr. Harry Foreman is Professor of Nuclear Medicine, School of Public Health, University of Minnesota; also Director of Center for Population Studies.

clides from the Monticello reactor that were 50 times more restrictive than those of the AEC. This set of regulations was paramount in the minds of many of the individuals from the State of Minnesota who attended the symposium.

In normal day-to-day operations, nuclear power plants are permitted by law to release radioactivity in the form of radioactive atoms to the environment in gaseous and liquid discharges. There are essentially two regulations concerned with these releases. The first regulation, which represents the primary standard, is the dosage that could be delivered to an individual or to the population-at-large. We have discussed this primary standard earlier in this book and have indicated that the standard is much too high, and that it would be a national tragedy for the population-at-large to be exposed to anything approaching this primary standard of 170 mr/yr.

The second regulation is a group of secondary standards. These are called the maximum permissible concentrations of the various radionuclides in air (MPC_a) and water (MPC_w.) that can be released outside of the restricted area of a nuclear reactor. The primary standard should be derivable from the secondary standards. But the secondary standards, the maximum permissible concentrations that are listed in Title 10 of the Code of Federal Regulations, do not permit this because they do not take into account the biological concentration mechanisms that actually take place in the environment between the release of the activity by the reactor and the eventual consumption of the contaminated foods by man.

Reactor releases would contaminate food

The MPCs that are tabulated in Title 10 of the Code of Federal Regulations apply only to the situation where individuals are breathing the contaminated air or drinking the contaminated water. They do not take into account the fact that the contaminated air and the contaminated water will result in the contamination of the foods consumed by man. This is an extremely important factor in terms of the dosage that would be received by man from reactor releases. For example, if the cesium-137 concentration in water were at the maximum permissible concentration allowed by the AEC, an individual could not eat even one pound of fish a year without exceeding the present radiation exposure guidelines.

The reason for this is that the cesium concentration in the fish will be a thousand times higher than the cesium concentration in the water. If the cesium-137 concentration in air were maintained, for just one day, at the concentration allowed by the AEC, a child drinking one liter of milk per day from cows on pasture would receive a dosage of 7 rad. This is 40 times the present exposure guideline for *one year!* Remember, this is for an air concentration at MPC for only one day, not for day after day. The same kind of calculation indicates that most of the MPCs are far too high. Tamplin presented this kind of information at the Minnesota symposium. He also indicated that the guideline dosage for exposure of the population was inappropriately too high and that no one should consider exposing the population to anything close to the guideline dosage.

The proponents of the nuclear power industry at this meeting, noticeably Dr. Merrill Eisenbud*, again tended to indicate that exposure at the guideline levels of radiation was not really significant.[44] They also indicated that the exposure from the nuclear power plants would be considerably lower than those of the guideline's. They indicated that individuals in the near vicinity of nuclear reactors would be exposed to no more than 5-10 mr/yr and that as individuals lived further and further away from the reactor, their exposure would drop off very rapidly from this 5-10 mr/yr value. Moreover, they indicated that the design objectives and the operation of existing power plants demonstrated that the actual releases of radioactivity from the power plants were no more than 1% of the releases allowed by the AEC's MPC values.

The message, then, that they presented to the individuals attending this symposium was that nuclear reactors would release

---

*Dr. Merrill Eisenbud is Professor of Environmental Medicine, New York University Medical Center.

radionuclides from the reactors at levels below those proposed by the State of Minnesota, which were 50 fold more restrictive than those of the AEC.

Consequently, the individuals attending the meeting repeatedly asked the question, "If reactors are only going to release the small amount of radioactivity that you indicate, then why are you so reluctant to make the guidelines more restrictive and adopt the Minnesota regulations?" This question was really never answered. Congressman Hosmer did state that if the standards were lowered he doubted if the reactors could operate safely. AEC Commissioner Thompson made essentially the same statement in testifying before the Joint Committee on Atomic Energy.[45]

In other words, the Joint Committee and the AEC state that the reactors will release radioactivity far below the guidelines but if you lower the guidelines they will not be able to operate safely. They seem to want to have their cake and eat it too. This is totally inconsistent.

What is the real reason?

If the reactors are to release less than 1/100th of the present allowable release rates, then why should the AEC and the Joint Committee be so reluctant to lower the standards by a factor of 50? The only conclusion that a reasonable person can come to is that the AEC and the Joint Committee do not believe that the reactors will be able to operate at these lower release rates. This becomes all the more confusing when you realize that the Northern States Power Company, which is most intimately concerned with the Monticello reactor in Minnesota, has indicated that it would be willing to comply with the State of Minnesota standards. To remove some of the confusion from this paradox, one might then conclude that the Joint Committee and the AEC are reluctant to lower the guidelines because this would not leave any room for the development of their other pet project, the Plowshare Program.

But this doesn't make very much sense because it is possible for a set of standards to be developed to apply specifically to the nuclear reactor program, and the Plowshare Program could then have a set of standards developed to apply directly to it. Accepting this, it must be concluded that the AEC and the Joint Committee do not have confidence that the nuclear reactor program will be able to operate when regulated by more restrictive standards. They must feel that these more restrictive standards will force the nuclear reactor industry out of business.

There are other aspects to the problem which were never really discussed at the Minnesota symposium that may give the AEC some cause for worry over more restrictive release rates on present-day reactors. These other aspects of the problem are the fuel reprocessing plants and the disposal of radioactive wastes. The fuel reprocessing and waste disposal aspects of the nuclear industry will be discussed in a subsequent chapter of this book. It suffices to say here that discussing the release rates from a successfully operating present-day nuclear reactor is only discussing a portion of the problem, that the major aspects of the problem reside in those subjects that were really not covered at the University of Minnesota symposium.

Another aspect of nuclear reactors that was largely glossed over at this symposium was the consequences of an accident. With respect to accidents, both minor and major, one important aspect of the nuclear power industry was brought out quite vividly by Harold Green, a law professor from George Washington University.[46] This was the requirement of the nuclear industry for the Price-Anderson Act.

Price-Anderson Act limits liability

The Price-Anderson Act is a unique piece of legislation. It limits the liability of a nuclear power reactor to some $560 million. This dollar limit is quite a concession since a study conducted by the AEC's Brookhaven National Laboratory indicated that an accident in a reactor could result in a total liability of some *$7 billion*. In other words, the Congress of the United States has limited the liability of these nuclear reactors in such a way that should a serious accident occur, the public would get back only 7 cents on every dollar of damage that was done.

Moreover, of this $560 million limit in liability, the private insurance carriers have underwritten less than $100 million. The U. S. Government stands behind the rest of the liability. It would seem that at the market place, where you have to put your money where your mouth is, the insurance underwriters have decided that they do not have sufficient confidence in this fledgling industry to underwrite even the Congressional limit of $560 million in liability.

A very important question arises from this consideration: "Would there be a nuclear power industry in this country if it were not for the Price-Anderson Act?" Most people feel that the nuclear power industry would not exist if the Price-Anderson Act did not co-exist with it. If the Act did not exist, the Government would probably be developing this reactor technology in remote parts of the country and the industry would be waiting to develop only when proven safety was established.

Another important point which Professor Green brought out at the Minnesota symposium was the cost to intervenors in reactor hearings. He estimated that, at a bare-bones minimum, it would cost the individual members of the public some $100,000 to fight a successful intervention against a proposed reactor. Moreover, he pointed out a number of inequities in the timing schedule for the intervenors and their lawyers to file petitions for intervention and to present their material before the licensing boards. In other words, he said that the reactor siting and licensing procedure was so organized that it would be almost impossible for those who opposed it to intervene successfully.

As a result of the symposium at the University of Minnesota, our apprehension concerning the safety of nuclear power reactors was increased. The reluctance of the AEC and the Joint Committee on Atomic Energy to make the standards relating to reactor releases more restrictive was a paradox since the AEC and members of the reactor industry stated that the releases would be very low. Moreover, the existence of the Price-Anderson Act and the reluctance of the insurance companies to underwrite even the limited liability which was set by Congress posed questions concerning the true safety of nuclear reactors.

Finally, the discussion of Professor Green concerning the difficulty in intervening in reactor siting and licensing procedures indicated that a nuclear juggernaut was moving across the land. If the nuclear industry, the AEC, and the JCAE were overstating the safety of nuclear power reactors, the situation looked grave indeed. We therefore began to look into the literature concerning the safety and operational aspects of the nuclear power reactors.

By the time the AAAS symposium arrived December 28, 1969, our review of the more general information concerning reactor safety had led us to the conclusion that the people of the United States were being subjected to a gigantic experiment. The unpredictable outcome of this experiment could be a severe tragedy occurring in some major metropolitan area. Consequently, the talk which Tamplin prepared for presentation at the AAAS meeting called for a moratorium on the construction of new nuclear power facilities until the uncertainties concerning the safety of nuclear power plants could be resolved.

Tamplin's moratorium suggestion is censored

It was this report that was severely censored by Dr. Roger Batzel, Associate Director of the Lawrence Radiation Laboratory for Chemistry and the Bio-Medical Division. In effect, Tamplin was informed that unless he removed the suggestion of a moratorium from that report, the laboratory would not fund his travel to the AAAS meeting; in fact, it would not allow him to prepare or have the report typed using the facilities of the laboratory. Dr. Batzel indicated that this kind of statement on the part of Tamplin represented irresponsibility.

Strangely enough, at that particular time there was such a bill, recommending a moratorium, introduced into the U. S. House of Representatives by Congressman Lester L. Wolff of New York.* Reluctantly, Tamplin removed that portion of the discus-

---
*The bill intended to provide a moratorium on the construction of new nuclear power reactors for a period of ten years.

sion from his paper for the AAAS meeting and at the AAAS symposium presented only the data related to the erroneous nature of the maximum permissible concentrations and the inappropriately high Federal Radiation Council guidelines for exposure of the population. We shall now present the full story as we saw it then with some additional information which has come to our attention subsequent to that period of time.

But before we go into the discussion of the nature of nuclear reactors, it is important to point out that one of the subjects discussed at the AAAS symposium was fossil fuel plants. The nature of the discussion concerning fossil fuel plants was that although today they belch noxious gases from their chimneys, it is totally within the ability of existing science and technology to clean up these emissions dramatically.

The development of the technology has been hampered considerably because of lack of funding for the necessary research and development. One of the reasons why this money may not have been available for the development of this technology was that some $400 million per year was pumped into the development of the nuclear reactor technology. The essence of the comments from the people discussing the fossil fuel plants was that we could have clean fossil fuel plants today.

These statements were highlighted in a talk by Carl E. Bagge, Vice Chairman of the Federal Power Commission:

> The research and development effort for atomic energy received over 84% of all the federal funds for energy R&D (research and development). It has also received approximately $3 billion of government expenditures in the past twenty years. Compared with this ambitious federal commitment to atomic energy, the amounts of money which have and are being allocated for the improvement of fossil fuel generation and for other fossil fuel energy research are ridiculously small.[47]

Moreover, at the University of Minnesota symposium, Dr. M. King Hubbert of the Department of Interior indicated, for example, that we have about a 200 year supply of coal in this country alone, in addition to our oil reserves. In other words, there is no real need to rush headlong into the rapid proliferation of nuclear power reactors because we have the potential for clean fossil fuel plants as a reasonable alternative to potentially unsafe nuclear power plants. We could wait for proven safety. Besides, as Power Commissioner Bagge and others have pointed out, fossil fuels will represent the backbone of our electrical power generation for a number of years to come. We should expend every effort to make fossil fuel plants as clean as possible.

What is wrong with nuclear power plants?

The normal day-to-day operations of a nuclear power plant are regulated by the standards tabulated in Title 10, Part 20, of the Code of Federal Regulations. These are the reactor regulations that are promulgated by the AEC and represent the basis for the licenses issued to the nuclear power plants. As we indicated in the early chapters of this book, the primary standard which sets the allowable level for the radiation exposure of the population-at-large is much too high. We estimate that if the population of the United States were exposed to this guideline there would be an additional 32,000 cancer deaths each year.

In addition to that, we estimate that the genetic consequences of this could be far greater, leading to an increase of between 150,000-1,500,000 additional deaths each year. In addition to these genetic deaths, there could be a 5-50% increase in such debilitating diseases as diabetes, schizophrenia, and rheumatoid arthritis. So far as the secondary standards are concerned, that is the maximum permissible concentrations in air and in water, we demonstrated in this chapter that these standards are essentially meaningless.

Therefore, one can state at the onset, one of the things that is wrong with nuclear power reactors is that the regulations which govern the normal day-to-day operations represent unacceptable guidelines, guidelines which would cause the public to pay too high a price for the benefit of nuclear-generated power. It may be that the nuclear power industry can meet far more restrictive standards than those promulgated by the AEC.

However, the reluctance of the AEC and the JCAE to change these standards causes us to have serious doubts as to whether or not the nuclear power industry will be able to meet more restrictive standards which actually represent acceptable standards for exposure to the population.

How radioactivity leaks through fuel rod pin holes

To a considerable extent, the amount of radioactivity released to the environment by an operating nuclear power reactor depends upon the integrity of the fuel rods in the reactor. The large reactors that are planned to be constructed and are being constructed in this country today have thousands of these fuel rods, which are inserted into the core of the reactor. As these fuel rods age, they develop small pin holes. The radioactivity which is generated within the fuel rods then leaks through these pin holes and into the water which is moderating the reactor. The reactor is not able to completely contain this water which bathes and moderates the fuel elements and contains the radioactivity which is leaked from the rods. Therefore, radioactivity-contaminated water accumulates within the reactor housing.

This radioactive waste water that had accumulated within the reactor is then metered out into the cooling water that is being returned to the river or to the ocean. Consequently, the degree below the maximum permissible concentrations that a given reactor will be able to operate depends upon the integrity of its fuel rods as well as the integrity of all the valves and nozzles and pipes that comprise the plumbing and cycling system of the reactor.

The reactors presently under construction are planned to operate for some 20 years. In addition to that, the plans are to change the fuel rods only once every two or three years. Moreover, these reactors are considerably larger than the reactors with which we have any experience to date. The combination of these factors indicates that we do not really know how reactors will operate as they begin to age and as their fuel rods begin to age. It may well be that the natural aging process of the reactor will cause it to creep up to the maximum permissible concentrations that are presently allowed by the AEC. They might even exceed those particular levels.

Since nuclear power reactors are being proposed at a rate which indicates they will be supplying a very substantial fraction of our future electrical power needs, we will be presented with a *fait accompli* in the future. If these reactors do not operate at their design specifications, it will be difficult to shut them down because we will need the power; and if we shut them down, sizeable sections of the country would experience periods of brown-out. We might, therefore, be forced to live with whatever radioactive emissions the reactors required. Once we have made a very sizeable commitment to nuclear-generated power, we must face the fact that we will be stuck with that commitment.

Safe disposal of old fuel rods a tremendous problem

There is yet another reason why the AEC is reluctant to make the guidelines for nuclear power reactors more restrictive, and this is the operation of the fuel reprocessing plant. When the fuel rods in a nuclear reactor have been in place for some two to three years, they will be removed from the reactor and taken to a fuel reprocessing plant. By fuel reprocessing, it is simply meant that the fuel rods are broken down and the remaining fissionable material is salvaged for the manufacture of additional fuel rods.

In this process of reclaiming the unspent fissionable material in the rods, the material is subjected to acid digestion. As a result, large quantities of liquid wastes are generated that contain relatively large amounts of radioactive fission products that were generated during the operation of the nuclear reactor. These large volumes of liquid waste, therefore, represent a substantial waste disposal problem.

The one commercially operable fuel reprocessing plant in the United States, the Nuclear Fuel Services Plant in West Valley, New York, is an example of this kind of operation. Although not operating at its full capacity, it is discharging large amounts of radioactivity into Cattaraugus Creek. As a consequence, downstream from the plant we calculate at the present time that if an individual consumed one pound of fish a week from this creek he would be exceeding the present radiation exposure guidelines.

With the proposed expansion of nuclear reactors, there will also have to be a commensurate expansion in the fuel reprocessing facilities. Consequently, the number of waterways which will be contaminated to a level commensurate with that associated with the Nuclear Fuel Services Plant could be numerous. In other words, although the nuclear power reactors themselves may release only small quantities of radioactivity into the environment, this will be more than compensated for by the operation of the fuel reprocessing plants. As a result, the present guidelines promulgated by the AEC may be just barely adequate and maybe not even that. We will discuss this tremendous radioactive waste disposal problem in more detail in the next chapter.

The reluctance of the Atomic Energy Commission, the Joint Committee on Atomic Energy, and the nuclear power industry to have more restrictive radiation protection guidelines strongly suggests that the present generation of nuclear power reactors will not be able to operate safely even in their normal day-to-day operations. This alone is sufficient to cause us to have serious apprehensions concerning the nuclear power industry.

Insurance firms reluctant to assume nuclear risks

But above and beyond the normal day-to-day operations there exists the possibility of an accident, either minor or major. As we indicated earlier, a study conducted by the Brookhaven National Laboratory indicated that a major accident within a nuclear power reactor could result in the deaths of several thousand individuals and cause some $7 billion in property damage. Because of this, the nuclear power industry in this country would not even exist if it were not for the Price-Anderson Act, which limits the liability of nuclear reactors to $560 million. Even then, the private insurance underwriters in this country will only underwrite some $100 million of this $560 million liability limit. The private insurance underwriters do not have that much confidence in this fledgling industry. An important question, then, is how safe are nuclear reactors?

To answer this question, let us quote from Walter H. Jordan, Assistant Director of the Oak Ridge National Laboratory and a member of one of the AEC's reactor safety boards. He states:

> The important question still remains: Have we succeeded in reducing the risk to a tolerable level, that is, something less than 1 chance in 10,000, that a reactor will have a serious accident in a year?
> Have we succeeded in reducing the hazard to such a low level? There is no way to prove it. We have accumulated so far some 100 reactor years of accident-free operation of commercial nuclear electric power stations in the U. S. This is a long way from 10,000 so it does not tell us much.
> The only way we will know what the odds really are is by continuing to accumulate experience in operating reactors. There is some risk but it is certainly worth it.[48]

How safe are nuclear reactors? Let us quote from a consulting engineer, Adolph Ackerman, of Madison, Wisconsin:

> As an independent consulting engineer I have been active for many years in alerting the engineering profession to its overriding responsibilities in design and construction of safe atomic power plants. The simple fact is that none of the atomic power plants currently in operation or under construction have been designed with the traditional concepts of engineering responsibility and ethical commitment for maximum public safety.[49]

How safe are nuclear power reactors? Let us quote from Robert L. Whitelaw, Professor of Law of the Virginia Polytechnic Institute and formerly Project Engineer for the design and construction of the power plant for the nuclear ship Savannah.

> I wish to endorse fully the principal argument advanced by A. J. Ackerman in his paper and, perhaps, strengthen the impact of his paper with this brief discussion.
> His principal argument has been confirmed by my own experience of the past fifteen years on nuclear projects and problems of

various kinds. This experience included preparing proposals and nuclear hazards evaluations on a variety of nuclear power plants, both commercial and military.

It has been my observation that, despite the enormous amount of meticulous detail which the ACRS regularly requires on every projected power plant to satisfy itself that there is no "credible accident" that can threaten the public (or even the operators)—and despite the volumes of paper and hours or presentations consumed on this topic, and no doubt well-intentioned—there is still by common consent an unwritten agreement to treat as "incredible" the most fearful of all nuclear accidents that can occur in any plant with a highly pressurized primary system. Such an accident is, of course, the explosive rupture of the primary vessel itself, which is ruled out of the list of credible accidents for the simple reason that there is no adequate answer short of putting the plant underground or inside a mountain, as Ackerman has pointed out.[50]

How safe are nuclear reactors? Let us quote the first chairman of the Atomic Energy Commission, David Lilienthal. He states:

> Once a bright hope, shared by all mankind, including myself, the rash proliferation of atomic power plants has become one of the ugliest clouds hanging over America.[51]

How safe are nuclear reactors? Let us quote from Dr. Edward Teller, often called the father of the hydrogen bomb, and one of the outstanding supporters of the AEC. Dr. Teller states:

> A single major mishap in a nuclear reactor could cause extreme damage, not because of the explosive force, but because of the radioactive contamination . . . So far, we have been extremely lucky . . . but with the spread of industrialization, with the greater number of simians monkeying around with things they do not completely understand, sooner or later a fool will prove greater than the proof even in a foolproof system.[52]

How safe are nuclear reactors? Let us quote from a letter of the AEC's Advisory Committee on Reactor Safeguards concerning a reactor planned for Midland, Michigan:

> The number of permanent residents within five miles of the

plant site was estimated to be 41,000 in 1968, mainly in the city of Midland and its environs.

The applicant has established criteria for, and has begun the formulation of a comprehensive emergency evacuation plan.[53]

In considering the safety of nuclear reactors, it is important to recognize that each nuclear reactor in this country is an experiment. Each reactor is different from all other reactors and whether or not it will operate and/or operate safely depends upon the outcome of the experiment. Before and subsequent to the granting of the construction permit the design of these reactors is modified. This is one of the reasons why nuclear power plants are costing more than originally estimated. For example, the cost for a plant in Fort Calhoun, Nebraska, is now twice the original estimate. Another consequence of these modifications is that the start-up date for nuclear power plants is usually much longer than was anticipated. For example, the Fort Calhoun plant will now be a year late in its start-up operation if nothing adverse occurs in the intervening period.

The hoped-for low cost of nuclear plants was, of course, one of the major reasons why many utility companies rushed in to buy these plants. The subsequent drastic increase in the cost of the plants and the delays in their construction, however, has cooled many utility companies' enthusiasm for nuclear plants.

Construction flaws are detected after operations begin

But the experimental phase of the nuclear reactors is not over when they go into their operational phase. After these reactors have been operating for a period of time, flaws in their construction are uncovered and expensive retrofits are often necessary. In the case of many of the first generation of plants, they had to be shut down and were really never operable. The new plants which are under construction and in operation today are also experiencing these errors in their experimental design.

Nucleonics Week published a fairly large discussion on the problems developed with furnace-sensitized stainless steel in critical areas of the reactors. This article indicates that trouble was encountered at Oyster Creek, Tarapur, Nine-Mile Point, and LaCrosse. These problems developed in furnace-sensitized stainless steel safe ends and other miscellaneous supports in the reactors.[54]

A somewhat similar problem developed in the Indian Point reactor on May 20. In this particular case, small pieces of material were found circulating in the cooling water. Since the reactors were constructed in order to meet critical power needs, it appears quite possible that, beginning in the year 1970, brownouts will occur as a result of their failure. Let us hope that none of the failures will be a serious accident.

To go on further concerning the safety of nuclear reactors, consider the unhappiness of the Advisory Committee on Reactor Safeguards with the AEC's reactor safety program and execution as discussed by *Nucleonics Week:*

> Reactor-safety research has been an ever-more-bludgeoned victim of the budget pressures on AEC, and the Advisory Committee on Reactor Safeguards is unhappy both about the diminishing supply of safety-research funds and about some of the decisions AEC is making on how to spend the funds . . .
>
> ACRS is dismayed over the constantly shrinking reactor-safety effort at AEC, by AEC's failure to follow through on many ACRS recommendations for safety r&d, by the sluggishness at AEC that makes it virtually impossible to get a speedy investigation of a specific problem that might arise during a project review, and by AEC's failure to schedule certain r&d—for whatever reason—in these specific areas . . .
>
> Water reactors. Despite recommendations of both ACRS and the regulatory staff for more r&d on fuel failure and partial or large scale core melting, "only small or modest efforts have been initiated thus far."[55]

These complaints by the AEC's Advisory Committee on Reactor Safeguards suggest that the present reactors and those under construction are even far more experimental than we might have imagined. The complaints on the part of the ACRS do little to allay our fears concerning the safety or reliability of nuclear power reactors.

How safe and reliable are nuclear power reactors? Unfortunately, it appears that no one really knows. The people of the United States are being asked to engage in a gigantic experiment. The stakes in the experiment may be extremely high, the losers paying off possibly even with their lives.

This is incredible when one considers that we have acceptable alternatives to nuclear power reactors. One such alternative that we have in this country is about a 200 year supply of coal. The following comments by Mr. Bagge (Vice Chairman of the Federal Power Commission) suggests very strongly that we have put our eggs in the wrong basket and that we should proceed forthwith to modify this mistake.

> The commitment by this industry to nuclear power generation also lies at the root of the power crisis. Stimulated by government policy, utility planners envisioned nuclear power as the answer to future expansion of their generating capacity and placed an inordinate amount of their eggs in the nuclear basket. And now the chickens have hatched and come home to roost.
>
> Although these vast nuclear generating complexes were welcome additions in the fight against air pollution, they created a new problem of thermal pollution which this industry for a while, insisted on characterizing as "thermal enrichment." Economy had also been one of the virtues of this mode of power generation. Now, however, the statistics from recently installed units were knocking earlier cost predictions out of the hat. The cost of skilled labor, quality control and stricter safety measures—all acted to skew investment curves beyond acceptable limits. New units did not become operational on schedule and suddenly the manufacturers were reporting that orders for new nuclear facilities had dropped to the level of the 1950s.
>
> The hard facts had to be faced—nuclear power generation was not the great panacea we had envisioned.[56]

We have moved too rapidly into the field of nuclear power reactors and it is high time that we begin to slow and even stop this pace. We should do this because there are inherent dangers in the present generation of nuclear power reactors. We simply do not yet know how to live safely with the peaceful atom.

But there is another compelling reason for slowing and even stopping the pace of development of the present nuclear power reactors, and that is that at the present rate of development, the reactors will rapidly consume the easily available uranium fuel upon which they depend. The continued existence of the nuclear power industry depends upon the development of the fast-breeder reactor. As a consequence, the AEC is now engaged in promoting the fast-breeder program at a pace that is not consistent with sanity. Sites are being selected all over the country for the construction of experimental fast-breeder reactors; but we know far less about the construction and operational safety of the fast-breeder reactor than we know about the present generation of nuclear power reactors. In other words, we are compounding an already potentially dangerous situation. Again we can quote from Commissioner Bagge's address:

> But there were more than merely economic factors which led to this crisis. Technology, which had so brilliantly brought forth nuclear generation, was at an impasse in developing the acclaimed salvation for the future needs of this country—the fast breeder reactor. After the inauspicious record and nearly catastrophic disappointment of the Enrico Fermi facility, the AEC and the industry found that their ambitious endeavor had fallen far short of its projected goals. It would be at least 1985 before the fast breeder, even with a sufficient commitment of funds at the earliest possible date, would have an impact on the Nation's power generation.

The fast-breeder reactor

In the present reactors which are in operation and being constructed, the major fuel source is uranium-235. When uranium-235 captures a neutron, it undergoes fission; that is, it splits into two smaller atoms and in the process also emits some additional neutrons. The energy which is released upon the splitting of the U-235 atom is then used to heat water which drives a steam turbine to generate the electric power.

The present reactors are moderated by water. By moderated it is meant that the neutrons which the U-235 releases upon undergoing fission are slowed down by the water. Now, the present nuclear reactors also contain U-238, and when the U-238 captures a neutron it is converted to plutonium-239. In the present reactors, because the neutrons are moderated or slowed down by the water in the reactor, the U-235 emits fewer neutrons than it would if it underwent fission as a result of absorbing faster neutrons. As a consequence, less Pu-239 is made in the present reactors than U-235 that is burned. There is a net consumption of fissionable material; considerably more fissionable material is consumed than is produced.

In the fast-breeder reactors, however, the water moderator is removed and the neutrons which are captured in this case by the U-235 are faster or higher energy neutrons. The net result of this is that the U-235, when it undergoes fission, produces more neutrons than it would when it undergoes fission as a result of the absorption of a slow neutron. This production of additional neutrons results in a net increase in the production of plutonium over the amount of U-235 that is consumed. That is why these reactors are called breeders; they produce more fissionable material than they burn. The operation of the fast-breeder reactor, therefore, is anticipated to produce large quantities of fuel to operate reactors fueled by Pu-239 rather than the present uranium-fueled reactors.

But the requirement for higher energy neutrons in order to produce a breeder reactor means that the reactor will necessarily have to operate at a higher temperature. The neutron moderator, the water, is therefore removed from these reactors and the reactors are cooled with liquid sodium. Sodium is a highly reactive metal and liquid sodium will explode upon contact with water or air. Moreover, the fast breeders, in order to increase their breeding capacity, have to be made more compact than the present generation of reactors. Therefore, they will contain much more fissionable material in a smaller volume.

The breeder reactor, therefore, concentrates the fissionable material into a smaller volume and operates at a significantly higher temperature. These two specifications for a fast-breeder reactor represent the most serious engineering complications for such systems. The major concern with these reactors is an accident that might result in the concentration of fissionable materials into small volumes wherein the chain reaction can proceed in an unmoderated fashion.

Fast-breeder reactor requires stringent control

Such an event could result in an extreme increase in temperature and a possible explosion. These explosions are not as tremendous as those which result from atomic bombs which are designed for this particular purpose, but nevertheless it is these potential explosions which represent the grave concern of nuclear-reactor designers. Consequently, the fast-breeder reactor places the most stringent requirements upon the control of the process to prevent over-heating and melting of the fuel materials. This is so because very small melts can result in the accumulation of critical masses and subsequent rapid elevations in temperatures—which is the process which produces the explosive force.

We can, in order to highlight the hazards associated with the fast-breeder reactor, again quote from Dr. Edward Teller, in a paper published in an issue of *Nuclear News*.

> For the fast breeder to work in its steady-state breeding condition you probably need something like half a ton of plutonium. In order that it should work economically in a sufficiently big power-producing unit, it probably needs quite a bit more than one ton of plutonium. I do not like the hazard involved. I suggested that nuclear reactors are a blessing because they are clean. They *are* clean as long as they function as planned, but if they malfunction in a massive manner, which can happen in principle, they can release enough fission products to kill a tremendous number of people.
>
> ... if you put together two tons of plutonium in a breeder, one-tenth of one percent of this material could become critical ... Although I believe it is possible to analyze the immediate consequences of an accident, I do not believe it is possible to analyze and foresee the secondary consequences. In an accident involving a plutonium reactor, a couple of tons of plutonium can melt. I don't think anybody can foresee where 1 or 2 or 5% of this plutonium will find itself and how it will get mixed with some

other material. A small fraction of the original charge can become a great hazard.⁵⁷

There is in existence at the present time only one large scale fast-breeder reactor. This is the Fermi reactor situated about 30 miles outside of Detroit. The Fermi reactor was the great hope of the nuclear power industry. On October 4, 1966, the Fermi reactor was being operated at one-tenth of its rated capacity when something occurred within it. The safety rods were inserted their full distance to shut down the reactor completely.

Once the safety rods had managed to halt the chain reaction and bring the reactor to a "safe" condition, the officials in charge of the plant were presented with a sizeable dilemma. They really didn't know how much damage had been done to the reactor. They weren't certain what they could do with the reactor in order to investigate the degree of damage that had been done. There was considerable anxiety about tampering with the reactor because this might cause the material in the reactor to relocate and form a critical mass which could then result in a substantial explosion. This explosion could have ruptured the reactor containment structure and sent lethal quantities of radioactivity on its way to Detroit.

Fermi accident exceeded maximum limits

Luckily, when they did begin to manipulate the reactor core, nothing serious happened. Everyone breathed a sigh of relief. But the final investigation of the accident at Fermi indicated that what had happened exceeded the maximum credible accident which was proposed in the hazard analysis report submitted prior to the construction of the reactor.*

Officials are again about to restart the Fermi reactor. In the process of reloading the fuel rods, they recently had a sodium explosion. For some reason, the AEC and the engineers who call this reactor "baby" are continuing their efforts to restart it.

---

*For a detailed discussion of this and other reactor accidents see Sheldon Novick, *The Careless Atom* (1970, pap.).

Considering its closeness to Detroit and its near catastrophic beginning, one wonders why.

Why are our technologists so arrogant and why are the citizens of Detroit so complacent? We see here the effect of the Gallant Knight.

As Federal Power Commissioner Bagge stated, we put the lion's share of our nation's research and development funds into the fission reactor and today we know that was a mistake. Because of the failure of nuclear reactors to meet their over-optimistic projections, fossil fuels will represent the backbone of the electrical power industry for at least the next two to three decades. It is essential that we spend the funds to make fossil fuel plants as pollution-free as possible.

Fusion energy—the promise of the future

Moreover, we should stop dotting the landscape with these experimental nuclear reactors. Such experimentation should be cautiously conducted in remote areas, not 30 miles from large population centers. Our coal reserves alone are more than sufficient to see us through a safe development period.

Many have expressed doubts as to whether fission reactors will ever represent an acceptable answer to our electrical power needs. The problems associated with disposing of their radioactive waste constitute a strong argument against them. The projected principal power source of the future is the fusion reactor. Some of the more recent estimates suggest that a successful fusion reactor can be developed within the same time period as that required for the fast-breeder reactor. In the presence of a fusion reactor the fast breeder would be obsolete. The present nuclear reactors and the fast breeders are looking more and more like a potentially dangerous, lame-duck technology. If these fission reactors aren't exploited quickly, they will most likely never be used.

Fusion power makes use of the nuclear reactions that are occurring in the sun. The nuclei of the lighter atoms, such as hydrogen isotopes, are fused together to form heavier atoms. By this process, large amounts of energy are released. Besides being intrinsically much safer, fusion reactors will produce millions of times less radioactivity than fission reactors.

Fusion-power research has been supported with some $25 million a year while fission-reactor research has received between $400 million and $500 million a year.

Three actions would appear to be necessary: (1) The construction of the experimental fission reactors should be stopped; (2) the pollution from the fossil fuel plants should be drastically reduced; and (3) fusion-reactor research should be given a much larger share of our national energy research and development dollars and manpower.

8 Undisposable radioactive waste

One of the major legacies of the nuclear age is radioactive waste. Discussions concerning the disposal of radioactive waste are misleading because these wastes are not disposable. They must be kept isolated from the environment and, therefore, "guardianship of nuclear waste" is a more meaningful concept. We are producing waste products that must be maintained in isolation from the environment for a thousand years or more. Guarding this radioactive garbage is one of the prices that future generations will have to pay, in addition to the genetic consequences they will suffer from the radioactivity which we are presently introducing into the environment, either deliberately or under the guise of waste disposal.

In the 1950s and early 1960s the great superpowers, the United States and the USSR, tested nuclear weapons in the atmosphere representing some 300-400 megatons TNT equivalent of nuclear fission-type bombs. The concern over the long-lived radioactive by-products, cesium-137 and strontium-90, resulting from these explosions, was enough to touch off a worldwide controversy. The cesium and strontium have introduced a biological hazard to the environment, and as a result, almost every living person carries both of these radionuclides in such tissues as muscle and bone. Our foodstuffs, grown on soil contaminated with both of these long-persisting nuclides, are regularly and essentially universally contaminated by them. The atmospheric nuclear test ban treaty has cut down the introduction of more of these poisons, at least for the two super-powers and the other signatories to this treaty. Clearly, the radioactivity from 300-400 megatons of nuclear fission bombs was worrisome to the world.

A large nuclear electric plant producing 1,000 megawatts of electrical power uses the same amount of uranium *in one year* as a 25 megaton uranium-fission bomb. And this means the production of strontium-90 and cesium-137 and other radioisotopes equivalent to that produced in such a 25 megaton bomb. As a consequence, those nuclear power plants, which are already on order, will produce 10 times as much radioactivity *each and every year* as was produced by all the superpower atmospheric weapon tests. By the year 2000, this is projected to grow to 100 times as much radioactivity as the weapon tests. Unless this material is isolated from the biosphere, we could, over a period of a few years or even each year, release more radioactivity to the environment than was released in all the weapon tests combined. Unfortunately, the evidence to date indicates that appreciable quantities of this reactor radioactivity will find its way into the environment and man.

"Everything but the squeal"

As we indicated earlier, the Gallant Knight of the nuclear industry is promoting not only its major product line but is also promoting its by-products. Consider the 1969 report of the Atomic Energy Commission, *The Nuclear Industry,* in which the following can be found:

Useful by-products from reactor waste

Fission products, such as strontium, cesium, and promethium, recovered during irradiated fuel processing operations, are already finding some useful commercial applications such as industrial thickness gauges, food irradiators, teletherapy units, as a power source in remote weather stations, etc. Others, such as xenon, krypton, rhodium, and palladium, are being considered for recovery because of their potential use in the electrical, jewelry, oil, and chemical industries. Possible markets for the expanded use of these materials in the near future offer many challenging opportunities.

Late in 1968, the AEC announced that the Richland Operations Office would seek expressions of interest from industry in the recovery of fission products rhodium, palladium, and technitium from the Hanford high level waste. Considerable interest was indicated by several firms and one, PPG Industries, is exploring the possibilities of recovering these fission products by a proprietary process, using a sample of the Hanford waste.

Of particular interest in the by-product category is neptunium, which is used as the target material in the production of plutonium-238. It is possible that at some future date there will be a very large demand for Pu-238, for use as a power source in our space program and also there could be large demands for the artificial heart program if it is successful. General Electric is offering to recover neptunium (as well as uranium and plutonium) from irradiated nuclear fuel at its chemical reprocessing plant being constructed at Morris, Illinois. Nuclear Fuel Services, Inc., New York State Atomic and Space Development Authority, and all the other companies with interests in the chemical reprocessing business are giving serious consideration to this and other isotopes for which a market and economic conditions justify recovery.

There are about 100 private firms that produce radioisotopes or convert them into products for medicine, science, and industry. Total sales of these companies are estimated at $53 million annually, consisting of about $8 million in basic radioisotope materials, $16 million in radiochemicals, $25 million in radiopharmaceuticals, and $4 million in radiation sources. In addition, sales of devices in which radioisotopes are employed total about $40 million a year. If the sales of products produced by radiation processing, auxiliary materials, and services related to radioisotope and radiation uses are included, the total commercial activity in the United States is at a level of several hundred million dollars annually.[58]

A very large fraction of these isotopes will eventually find their way into the environment via sewers and garbage heaps. Much of this radioactivity will, as has already occurred, find its way into the environment through transportation accidents. Moreover, some shipments will simply be lost in transit and some will be misplaced and forgotten.

Some of the broken, misplaced and forgotten radioactive

products will produce serious immediate consequences. As an example, we can cite the case of a young Mexican boy who found a cobalt-60 source. The source, highly radioactive, looked to him like a metallic marble and he put it in his pocket. The radioactivity subsequently made him ill and his mother put him to bed. She put the "marble" in a drawer in the kitchen. As a consequence, both the mother and the boy's little sister became ill and the maternal grandmother came to care for them. The final result was that all four died.

Other such tragic examples are available. However, it is important to recognize that the hidden genetic consequences of the introduction of this radioactivity to man's environment will pale these examples by comparison. Undoubtedly, there are some necessary applications of radioisotopes, but one thing is certain—we do not need a "gung-ho" radioisotope industry, and the industry which we presently have needs to be more strictly regulated.

Fuel reprocessing plants and waste "Disposal"

Fuel reprocessing plants take the spent fuel rods from reactors and reclaim the fissionable material. The AEC has several fuel reprocessing and waste storage and disposal sites that we shall discuss subsequently.

At the present time, there is one private fuel reprocessing plant, the Nuclear Fuel Services plant in West Valley, New York. This plant is regulated by the same set of regulations that applies to reactors. As we indicated in the chapter on nuclear reactors, these regulations are a travesty on the public health. Moreover, we indicated in that chapter that an individual would be exposed to the guideline dosage if he ate only one pound of fish per week from Cattaraugus Creek into which the West Valley plant's wastes are released. A U. S. Health, Education and Welfare Department publication discusses the plant. In it, we find that "suckers are taken from Cattaraugus Creek for food, especially in the springtime. In addition, there may be a practice of grinding the flesh and bone for fish burgers."[59]

Nuclear Fuel Services is also licensed by the AEC for waste "disposal." As a result, the company has a burial site on the West Valley compound. The same publication shows that 10-15% of the radioactivity in the creek is coming from this so-called burial site.[60]

Even low discharge rates can contaminate water

The Nuclear Fuel Services operation gives us little reason to become complacent about the nuclear industry. Even if the reactors maintain low discharge rates, the West Valley story indicates that these fuel reprocessing plants and waste burial sites may well bring our ground water and rivers to the limit of the regulations. We can get little solace from the AEC on this point. Their 1969 publication, *The Nuclear Industry,* states:

> Intermediate level liquid wastes is a term applicable only to radioactive liquids in a processing status which must eventually be treated to produce a low level liquid waste (which can be released), and a high level waste concentrate (which must be isolated from the biosphere).
> Low level liquid wastes are defined as those wastes which, after suitable treatment, can be discharged to the biosphere without exposing people to concentrations in excess of those permitted by AEC regulations.
> Wastes generated in the cold or pre-irradiation phase of the fuel cycle (from the mine to the reactor), as well as wastes resulting from research laboratories and from medical and industrial applications of radioisotopes, are generally considered as low level or low hazard potential wastes.[61]

Considering the AEC regulations, low hazard potential really means—fools rush in where wise men fear to tread. The West Valley plant indicates that commercial operations represent a serious hazard. The National Academy of Sciences-National Research Council indicates that the AEC operations are not much better. Subsequent to the Minnesota symposium (October 1969), "Nuclear Power and the Public," Dr. Harry Foreman, the program chairman, refused to allow Dr. M. King Hubbert to insert the following into his text:

HUBBERT, *"Industrial Energy Resources"*

In view of the public concern over the question of environmental contamination by atomic wastes from nuclear reactors manifested at this meeting, it is pertinent to state that during 1965, at the request of the Atomic Energy Commission, the Committee on Geologic Aspects of Radioactive Waste Disposal of the Division of Earth Sciences, National Academy of Sciences-National Research Council, made a final review of the waste-disposal practices at the following establishments of the Atomic Energy Commission:

1. Savannah River Laboratory, South Carolina
2. Oak Ridge National Laboratory, Oak Ridge, Tennessee
3. Carey Salt Company mines, Hutchinson and Lyons, Kansas, to witness storage of high-level wastes in accordance with the same committee's earlier recommendations
4. National Reactor Test Station, Arco, Idaho
5. Hanford Atomic Products Operations on the Columbia River in Washington.

The report of this committee, dated May 1966 and comprising 92 pages, was submitted to the Atomic Energy Commission. Notwithstanding the fact that all earlier reports of this committee had been released to the public, and despite repeated requests from two successive chairmen of the Earth Sciences Division, release of this report by the Atomic Energy Commission has been persistently refused.

This represents a particular instance—one of my personal acquaintance—of the AEC practice, described so effectively by Dr. Harold P. Green in his paper before this symposium, of withholding from the public, or "sweeping under the rug," information that is unfavorable to the "cosmetising" treatment to which most AEC documents are subjected. While only the Atomic Energy Commission has the authority to release this report, it appears that the time has come when a concerned public, as well as Chairman John E. Moss of the Government Operations Subcommittee on Government Information, should at least become aware of its existence.[62]

Dr. Earl Cook* reported further revealing information:

I have had some experience with "the parochial interests in

---

*Dr. Earl F. Cook is Associate Dean, Geosciences, College of Geosciences, Texas A&M University.

atomic energy." I was staff officer for the National Academy of Science Committee on the Geological Aspects of Radioactive Waste Disposal, whose report to the AEC on storage and disposal policy and practice at the main AEC waste-producing installations was never released by the Commission, despite repeated attempts by the Academy's Division of Earth Sciences to have the report made public. The committee was discharged, and a new one—with no overlap of membership—was appointed. Informal inquiry to the AEC's Idaho Falls office about the Committee's report by one of my Idaho newspaper friends brought the reply that the report contained "serious scientific errors" which would embarrass the Committee members were it to be made public. In other words, the AEC is protecting, not inself, but the Committee, in refusing to release the report! Does that sound familiar?[63]

We were appalled to learn that not only was the AEC able to silence the National Academy, but it appears that it was even able to manipulate its committees. In response to pressure from Senators Church and Muskie, the AEC has finally released this report. As you might imagine, it is critical of the AEC's waste storage and disposal practices.

The Committee report contained two important conclusions: (1) None of the existing AEC disposal installations are in a satisfactory geologic location; and (2) present practices of disposing of intermediate and low level liquid waste and all manner of solid waste directly into the ground would, in the long run, lead to serious fouling of man's environment.

The 1969 report, *The Nuclear Industry,* quoted above, indicates that the AEC has not heeded the advice of the Committee. The Nuclear Fuel Services plant indicates that the time for taking the Committee's recommendations seriously has already passed. We have no adequate means of containing these low- and intermediate-level radioactive wastes and the proliferation of nuclear reactors is only going to compound an already serious problem.

9 Plutonium: Public health and technological arrogance

Plutonium is an element that was virtually nonexistent in the earth's natural crust. The present worldwide inventory of plutonium is, for the most part, man-made. By far the major use of plutonium at the present time is in the manufacture of nuclear bombs. Consequently, although it was not intended by its discoverers, the name plutonium is quite apt for this element: plutonium, the element of the Lord of Hell.

Plutonium has several isotopes, the most important being plutonium-239, which is used in the manufacture of nuclear bombs. In addition, Pu-239 is planned to be the major fissionable fuel in the future as a result of the development of the fast-breeder reactors.

Pu-239 is a radioactive isotope of extremely long half-life, 24,000 years; hence, its radioactivity will remain undiminished within human time scales. The other isotope of plutonium that is of interest is Pu-238. This isotope has a much shorter half-life, about 80 years. Interest in it has developed because it can be used in small power supplies for the production of power to drive such things as mechanical heart pumps and pacemakers that adjust the heart beat of humans. It is presently used in auxiliary power systems for space vehicles. The space vehicle application of Pu-238 will be discussed later in this chapter.

The cancer-inducing potential of plutonium is well known. An amount as small as 1 ten-millionth of an ounce injected under the skin of mice has caused cancer. A similar amount injected into the blood stream of dogs induces a substantial incidence of bone cancer. Fortunately, the body maintains a relatively effective barrier against the entry of plutonium into the blood stream. Also, because of the short-range in tissue of the radiation emitted by plutonium isotopes, the radiation from plutonium deposited on the surface of human skin does not usually penetrate to the depth of any sensitive tissue. Unfortunately, the lung is quite vulnerable to plutonium.

The vulnerability of the lung to plutonium results from the fact that when plutonium metal is exposed to air, it ignites spontaneously. As it burns, it forms numerous very small particles of plutonium oxide. When these particles of plutonium oxide are inhaled, they are deposited in the very deep portions of the lung. Tiny plutonium dioxide particles remain immobilized in the deep portion of the lung for hundreds of days, and during this time their radiation is able to affect the cancer-sensitive cells of the lung.

Particles of plutonium oxide represent an intense source of radiation. However, due to the nature of the radiation which they emit, they will only irradiate a small volume of tissue. For example, if the plutonium oxide contained in a tiny particle, four one-hundred-thousandths of an inch in diameter, were distributed uniformly throughout the human lung, the plutonium would deliver a dosage of 0.0002 rem per year to the lung tissue. On the other hand, if this plutonium oxide were contained in a single particle, the average dosage to that tissue which would be irradiated would be 500 rem per year and, to the closest 20 alveoli (lung sacs), the dose would exceed 3,000 rem per year. These dosages are a million to 10-million times larger than the dosage that would be delivered if this plutonium were distributed uniformly throughout the entire lung.

Plutonium tolerance level may err greatly

Thus, particles of plutonium oxide result in an intense, but localized tissue irradiation. Particles, therefore, result in an extremely nonhomogeneous distribution of the dosage to the lung. A small portion of the lung tissue is subjected to extremely high dosages of irradiation from a particle while the rest of the lung tissue is totally unaffected by the particle's radiation.

The International Commission on Radiological Protection has recommended a tolerance level for the exposure of human lungs to plutonium. However, the tolerance level relates to the uniform distribution of plutonium throughout the entire lung, and the ICRP indicates that this may be greatly in error when applied to particulates of plutonium oxide which result in this nonhomogeneous distribution of the dose. The Commission states:

> In the meantime, there is no clear evidence to show whether, with the given mean absorbed dose, the biological risk associated with a nonhomogeneous distribution is greater or less than the risk resulting from a more diffuse distribution of that dose in the lung.[64]

The Commission is effectively saying that there is no guidance with respect to the risk from the nonhomogeneous exposure of the lung by particles of plutonium oxide. Hence, the maximum permissible lung burden is meaningless for plutonium particles and so is the maximum permissible air concentration which is derived from it. By now, this far along in this book, the reader will undoubtedly not be surprised to discover that regardless of the statement of the ICRP, the ever-optimistic AEC has adopted the plutonium standard recommended by the ICRP for uniform lung irradiation to apply to particulates of plutonium oxide.

Some two years ago our colleague, Donald Geesaman,* began to review the literature concerned with the hazard of these highly radioactive particles in the deep portions of the lung. He was prompted to begin this study as a result of questions concerning the safety of space nuclear systems and questions asked by the directors of the Livermore laboratory concerning the weapons program.

His review of the literature pointed up a very sobering fact:

*Donald P. Geesaman is a research scientist at Bio-Medical Division, Lawrence Radiation Laboratory.

the experimental data indicated that when small portions of tissue were exposed to extremely high dosages of radiation, cancer was an almost inevitable result.[65] In other words, the experimental animal data suggested that irradiation by particulates of plutonium oxide represented a unique carcinogenic hazard. His analysis indicated that exposure at the maximum permissible lung burden for plutonium oxide, if the exposure occurred as particulates of plutonium oxide, was at least 100 times worse than imagined.

The studies which Geesaman was able to use to arrive at this estimation of the risk associated with plutonium oxide particles were not experiments dealing with the irradiation of the lung by plutonium oxide particles. Therefore, the results of his analysis were only capable of indicating that there was a very strong possibility that the effects of plutonium oxide particles were far worse than the uniformly distributed dose to the lung.

Experiment fails to consider low-dosage effect

The one experiment funded by the AEC that related directly to the exposure of plutonium oxide in the lungs was conducted at the Hanford facility of the AEC by Dr. W. J. Bair.* However, the design of the study proved to be its major failure. In the typical over-optimistic attitude of the AEC's Division of Biology and Medicine, the assumption appears to have been made that exposure at the allowable plutonium guidelines would result in an insignificant damage to the exposed individual. It appears that, using this assumption, the experiment on beagle dogs at Hanford was, therefore, conducted at dosage levels 100 or more times that of the maximum permissible concentration. The assumption seems to have been that one had to expose the animals considerably above the guidelines to be able to observe any biological effect of the radiation.

The results of the experiment undoubtedly came as a tremendous surprise to the AEC. Essentially all of the animals

*Dr. William J. Bair is associated with Pacific Northwest Laboratories, Batelle Memorial Institute.

that survived the first few years of the experiment developed lung cancer. The experiment tells us nothing about exposure to lower dosages. All the experiment indicates is that the AEC, in funding the experiment, was over-optimistic concerning the potential effects of plutonium oxide in the lung.

The AEC is reluctant to accept the analysis of Donald Geesaman. It chooses to apply the ICRP's recommendations for the uniform radiation of human lungs by plutonium at such installations as the Dow Chemical-operated Rocky Flats plutonium facility near Denver. The AEC uses this recommendation and applies it to the radiation from particulates of plutonium oxide. Geesaman does not really stand alone in his analysis, although he is the only one who has given a quantitative estimate of the enhanced risk from the particles. Dr. Bair, who performed the beagle plutonium experiment, and his colleagues, in a paper given in October 1969 at an Oak Ridge symposium on inhalation carcinogenesis stated that "Nonuniform irradiation of the lung from deposited radioactive particulates is clearly more carcinogenic than uniform exposure (on a total-lung basis) . . ."[66]

We may also quote from Dr. K. Z. Morgan's testimony in January 1970 before the Joint Committee on Atomic Energy of the U.S. Congress. Dr. Morgan is one of the United States' two members to the Main Committee of the International Commission on Radiological Protection. He has been a member of that committee longer than anyone else who is on it, and he is also Director of the Health Physics Division of the Oak Ridge National Laboratory. Dr. Morgan stated:

> There are many things about radiation exposure we do not understand, and there will continue to be uncertainties until health physics can provide a coherent theory of radiation damage. This is why some of the basic research studies of the U.S. AEC are so important. D. P. Geesaman and Tamplin have pointed out recently the problems of plutonium-239 particles and the uncertainty of the risk to a man who carries such a particle of high specific activity in his lungs.[67]

In a talk presented at the University of Colorado on April 19, 1970, Donald Geesaman stated:

Finally, I would like to describe the problem in a larger context. By the year 2000, plutonium-239 has been conjectured to be a major energy source. Commercial production is projected at 30 tons per year by 1980, in excess of 100 tons per year by 2000. Plutonium contamination is not an academic question. Unless fusion reactor feasibility is demonstrated in the near future, the commitment will be made to liquid metal fast-breeder reactors fueled by plutonium. Since fusion reactors are presently speculative, the decision for liquid metal fast breeders should be anticipated and plutonium should be considered as a major pollutant of remarkable toxicity and persistence. Considering the enormous economic inertia involved in the commitment, it is imperative that public health aspects be carefully and honestly defined prior to active promotion of the industry. To live sanely with plutonium, one must appreciate the potential magnitude of the risk, and be able to monitor against all significant hazards.

An indeterminate amount of plutonium has gone off site at a major facility [the Dow Chemical Rocky Flats plant] 10 miles upwind from a metropolitan area—Denver, Colorado. [The loss was unnoticed.] The origin is somewhat speculative as is the ultimate deposition.

The health and safety of public and workers are protected by a set of standards for plutonium acknowledged to be meaningless.

Such things make a travesty of public health, and raise serious questions about a hurried acceptance of nuclear energy.[68]

Space Nuclear Auxiliary Power (SNAP)

The space nuclear auxiliary power systems with which this section is concerned are those which contain plutonium-238. In these power systems the radioactive decay of the Pu-238 produces heat which is then converted into electrical energy. One of the space systems, which was designated SNAP 9A, burned-up in the atmosphere in April 1964. Subsequently, tiny particles of plutonium-238 oxide were distributed over the surface of the earth where they were inhaled to contaminate human lungs. The event caused President Johnson to insist that any subsequent SNAP reactors which were fired into space should be reviewed by a number of different boards composed of experts from the AEC, the Department of Defense, and National Aeronautics and Space Administration (NASA). As we shall see, regardless of the President's intent, these review boards represented nothing more than a facade. Tamplin became a member of the AEC's Division of Biological and Medicine Committee on Space Radiological Safety Matters.

The Martin Marietta Company study

The first meeting of this committee which Tamplin attended (June 1967) was related to the safety of a system designated as SNAP-19, which was to be launched along with a meteorological satellite from Vandenburg Air Force Base in California. The SNAP-19 was to supply auxiliary power to operate the meteorological satellite. At this meeting we were apprised of a rather extensive safety study that had been performed by the Martin Marietta Company, a participant in aerospace research and development. As part of this overall safety analysis, the Martin Marietta Company indicated, for example, that for a particular type of accident which might occur in association with the launch of this vehicle, the burn-up of the SNAP reactor in the atmosphere could result in some 40,000 fatal cases of lung cancer.

Tamplin's impression was, and still is, that the Martin Marietta Company had done quite a credible job on the safety analysis. He found fault with only two aspects of their analysis. One of the concerns that he had about their analysis was the biological hazard associated with the plutonium oxide particles. The other concern was with the reliability figures which purported to assess the probability of various types of accidents occurring.

In the analysis, probabilities of various types of accidents associated with the launch and insertion of the space vehicle into orbit were given as about 1 in 1000. It was quite apparent to Tamplin that the space program, in its entirety, had never developed enough experience with the firing of various types of rocket missiles to come to the conclusion that any particular event would have a probability of occurrence as low as 1 in 1000. Such probabilities can only be derived from experiments in which several thousand such launches have been made. Therefore, the estimated probabilities associated with various types of accidents in the launch and insertion sequence were a figment of analytical maneuvering, and as we shall see subsequently, these probabilities certainly had no relationship to the realities of the success of SNAP systems in the launch and insertion process.

Judgments should be based on consequences not probabilities

Tamplin contended at this particular meeting that one should ignore these probabilities, that one should look at the worst possible accident that could occur, and the entire safety analysis and judgment concerning the advisability of firing this system into space should be based upon the consequences of that particular kind of accident occurring. He indicated clearly that he had no confidence in their probability estimates concerning the possibility of the accident happening, and that, therefore, the only rational approach was to base one's judgments upon the consequences if such an accident occurred.

At the time of this meeting, Tamplin had not looked into the biological consequences of the inhalation of particles of Pu-238 or various other highly radioactive particles. In the course of the meeting individuals such as Dr. Wright Langham, of the Los Alamos Scientific Laboratory, presented their estimation of the hazard associated with inhalation of these highly radioactive particles.

One of the things that developed, much to Tamplin's amazement, was that the International Committee on Radiological Protection had given no guidance whatsoever with respect to the hazards associated with these particles. Moreover, in its reports, the ICRP indicated that it did not know whether the hazard was greater or less than the hazard associated with the uniform irradiation of the entire lung. An even greater source of amazement was that the major thrust of the individuals who presented data and arguments at this SNAP meeting was to indicate that the hazard from these plutonium oxide particles was significantly less than the uniform irradiation of the lung.

Plutonium: Public health and technological arrogance 185

Subsequent to this meeting Tamplin returned to the laboratory and requested Donald Geesaman to investigate the hazard of these particles. His analysis of the risk from these particles was presented above. As a result of Geesaman's study, we feel that the hazard from these particles is far worse than the hazard from the uniform irradiation of the lung, by as much as 100-fold.

The major thrust of the members of the AEC's DBM Committee was to discredit the biological risk aspects of the Martin Marietta safety analysis. A preliminary draft of the Committee's report concerning the safety of the SNAP-19 system was sent to Tamplin. In response to this draft he made the following statement to Dr. H. D. Bruner,* the Committee chairman:

> I am returning the draft of the SNAP-19 Committee report. I have several specific comments and one general comment. My general comment is that in reading the report, I felt that it gave the impression that these 35,000 curies were not of any real consequence, and arguments to the contrary, including Martin Marietta's estimates, were of little merit. I do not think that the existing scientific evidence is sufficient to warrant the creation of this impression. . . .[69]

The final draft of the Committee ignored Tamplin's comments, it refuted, degraded, and down-graded the Martin Marietta estimate of the risk, it implied that the overall probability of accident was so low that the biological hazard was, therefore, minimal and that the shot could be carried out as proposed.

The various committees decided that the proposed firing of the meteorological satellite with the SNAP-19 reactor was a reasonably safe operation to undertake; therefore, the system was launched from the Vandenburg Air Force Base. But quite quickly all of the masterfully calculated probabilities of accidents became nothing but figments of imagination because the rocket had to be blown up. The SNAP reactor fell into the waters off the San Clemente Islands near Santa Barbara, California.

The next meeting held by the AEC-DBM Space Nuclear Safety Committee relating to SNAP reactor systems was for the

*Dr. Harry D. Bruner is Assistant Director for Medical Research, Division of Biology and Medicine, AEC.

systems called SNAP-27 (in August 1968). These systems were to be placed aboard the moon landing vehicle in the Apollo portion of the space program. Again the safety analysis, performed this time by General Electric, indicated that in certain accident situations with these SNAP-27 systems, very sizeable numbers of fatal lung cancers could be induced in the humans on the earth.

Geesaman's conclusions get a cool reception

By this time Donald Geesaman had spent considerable effort in analyzing the potential risk of plutonium oxide particles in the lungs, and we had come to the conclusion that the hazard associated with these particles was 100 to 1,000 times worse than was imagined or used in the Martin Marietta analysis for SNAP-19. Geesaman presented his analysis and conclusions to the DBM Safety Committee. The best that can be said is that he was listened to with deference. The members of the Committee were unmoving in their opinion that the hazards of these particles were considerably less than that of the uniformly irradiated lung.

Again, in association with the potential lung cancers that could be induced, the General Electric Company included the probability of the various accidents occurring. We argued, as we had before, that experience with these space systems was not sufficient to have derived the rather optimistic probabilities that were associated with their potential failures. We stated that the probability should not be considered in the safety analysis, that one should consider that the accident could happen, and then evaluate the possible consequences. Judgment concerning the safety of the situation should be based upon that worst-case possibility. We were told not to worry about it; that although SNAP-19 did fall into the waters off Santa Barbara, we should recognize that we were now discussing man-rated systems and that the reliability of these systems was far greater than the reliability of the weather satellite system. We were told that such accidents had probabilities much less than 1 in 1000 in this particular case.

Plutonium: Public health and technological arrogance 187

We persisted in our argument that these probabilities of failure should not be part of the safety analysis, that we simply had insufficient knowledge about the systems to be able to quote such high reliability. Moreover, we argued that, in this particular case, an acceptable alternative to the procedure which was being followed with the SNAP-27 systems was available. The SNAP-27 systems were to be attached to the Lunar Excursion Module and, therefore, they would suffer the same fate as the Lunar Excursion Module (LEM). We suggested that they should put the SNAP reactor in the Command Module with the astronauts. We argued that since so many millions of dollars had been spent in trying to protect the lives of these three volunteers, who chose of their own will to sit on top of a Saturn rocket, it would be worth considering the expenditure of a small amount of money to protect the lives of several thousand individuals who might develop fatal lung cancer in case the mission aborted.

How SNAP accident probabilities misfired

We insisted that the SNAP reactor should be put in the Command Module where it could be recovered along with the astronauts. The promoters of the project indicated that they were too far along in the design of the entire system to consider this as a possibility and that really we shouldn't worry because this system was so terribly reliable; that the chances of these accidents were extremely remote indeed—so they said.

The final Committee report on SNAP-27 was submitted to the AEC. It reflected the Committee's confidence in the reliability of these systems and also its confidence that the hazard of plutonium-238 particles was less than that of the uniformly distributed dose to the lung. As a consequence of receiving this final report, Tamplin wrote again to Dr. H. D. Bruner, chairman of the Committee, indicating that he could not concur with this report, and that his name should be withdrawn from it. Further, if the report had been sent to anyone, he requested the AEC to write letters to those individuals indicating that his name should be withdrawn because he simply could not concur. Moreover, Tamplin indicated that since the Committee's conclusions were always so divergent from his own, he felt it incumbent upon himself to resign from it. Tamplin could only feel that his presence at the Committee meetings merely lent credence to the facade.

The first Apollo moon landing did not contain a SNAP reactor. The first SNAP reactor was included on the LEM in the second moon landing. This SNAP reactor is now on the moon and operating. Apollo 13, the third moon landing, also contained a SNAP-27 reactor. Of course, the mission of Apollo 13 was aborted, and the LEM which contained the SNAP-27 system re-entered the atmosphere.

Tamplin, therefore, had sat on committees which reviewed the safety of three SNAP reactors. He had been informed that the probabilities of failure and accidents were in the neighborhood of, and less than, 1 in a 1000. Of the three SNAP reactors considered, two of them resulted in accidents. Two out of three failures, while the probability of such failures was supposedly less than 1 in a 1000! Arrogance—this is arrogance in the extreme!

Moon landing was an 'engineering disaster'

It is important to point out one of the fallacies of the public's attitude toward the space program. The press has given the space program a great deal of play and an overabundance of praise. The landing of man on the moon has been likened in significance to the birth of Christ.

The truth of the matter is that the space program, the landing on the moon, was for the most part based upon technology that was available in 1959. The only reason we went to the moon was because the American public pumped $4 billion a year into that program.

If one looks realistically at the space program in terms of the amount of money that was pumped into it, it was an engineering disaster. We must remember that three Apollo astronauts were incinerated on the pad at Kennedy Space Flight Center, and that Apollo 13 narrowly missed killing three others. But the "success" of the space program has inflated the public's image of American science and technology.

Many individuals feel that if we get into difficulty with our environment, our science and technology can save us. Don't you believe it! Our scientists and technologists are not even able to guarantee that they can perform the feats they have on the drawing boards today. We should all consider carefully that science and scientists have very few answers and that science and scientists, for the most part, do not even know what are the appropriate questions to ask. So far as the success of our technology is concerned, one should consider the fact that the Golden Gate Bridge is simply lucky to be standing across the San Francisco Bay. An engineer who helped to design the Golden Gate Bridge subsequently designed a similar type bridge in the State of Washington. That bridge was literally blown down by the wind.

10 The nuclear weapons program

The development of nuclear explosive devices for use in warfare has been the major activity of the AEC. It is worthwhile to retrace the course of this nation's policy with respect to nuclear weapons and to relate this to the development of nuclear weapon programs coordinated between the AEC and the Department of Defense. Much the same as with the peaceful applications of nuclear energy, the proponents of nuclear weapons have proceeded, in a Madison Avenue approach, to push for the continued development and expansion of weapon systems. To a considerable extent, a demand for various nuclear weapon systems has been created in a manner quite similar to the demand for electricity which the power companies have created. Once these demands have been created, they are then stated to be needs.

At the end of World War II we had a monopoly on nuclear weapons and in the intervening years we proceeded at a comparatively slow pace in the development of our nuclear warfare capability. However, it soon became apparent that the Russians could and actually were developing nuclear weapons of their own. Consequently, we moved into the crash program to develop the so-called hydrogen or thermonuclear bomb. It was as part of this program that the Lawrence Radiation Laboratory in Livermore was created.

History now tells us that both the United States and the USSR developed thermonuclear weapons, that both nations proceeded from the comparatively small 20 kiloton atomic bombs that were dropped upon Hiroshima and Nagasaki into the development of multimegaton nuclear bombs. Thus, it developed that the two major protagonists on the world scene both had a devastating capability in terms of nuclear weapons. The world, particularly the United States and the USSR, had arrived at an unthinkable situation: each could destroy the other but was certain to be destroyed in return. This condition of unthinkability came about during the Eisenhower administration..

By the late fifties, it became apparent that nuclear power politics had become part of the cold war. The Eisenhower administration introduced what one must consider to be the least irrational attitude towards nuclear weapons, namely, that of massive retaliation. The policy which the United States adopted was that we would never use our nuclear weapons except in retaliation to a nuclear attack by another nation. At that particular time this meant the Soviet Union. The massive retaliation philosophy did not mean that one nation would win and another nation would lose. The basic concept behind the massive retaliation philosophy was, "Let's face it, nobody has anything to gain by engaging in a nuclear war. Not only is there nothing to gain by engaging in a nuclear war; there is practically everything to be lost." That was a realistic statement of the situation and it still is.

No rational policy possible for use of nuclear weapons

The USSR philosophy towards nuclear weapons has never really been clear to the people of the United States, although it has often been implied that the Russians would use their weapons on the United States first. The Eisenhower philosophy was that we would never use our weapons first; we would only use them in retaliation, and our retaliation would be massive. This could be considered to be a rather irrational, childish policy, but one must face the fact that this is the least irrational of all policies concerned with the use of nuclear weapons. There is no rational policy.

It was probably because this policy was so childlike that

individuals began to question it. However, even today it does not seem that the average man in the street has ever questioned the policy in any substantial way. The facts seem quite obvious to the average person and the limited degree of irrationality is accepted as the best approach towards the problem of nuclear weapons.

But it is easy to see how this approach would not be well received by the military mind or the strategic thinker. Therefore, military types and scientists associated with the military, most notably Herman Kahn, currently the director of Hudson Institute, began to conceive of devious strategies that might be employed in order to promote nuclear war. Mr. Kahn and others therefore introduced into the dialogue the concepts of first-strike capability and civil defense and the devious concept of nuclear blackmail. In so doing, they began to lay the groundwork for progressively more irrational approaches towards nuclear war and stimulated the arms race to its present position.

As a consequence of activities of individuals such as Herman Kahn, the thinkability of nuclear war was pressed upon the American public, and the great civil defense debate began. Nuclear war games were being played, the dead were being counted, and nuclear strategies were being developed. An Office of Civilian Defense was established and we were off towards a new era of formulating a nuclear weapon policy for this nation. A number of fallout shelter companies came into being. Governor Nelson Rockefeller spent some time sitting in a fallout shelter in public view. A great deal of effort was put into the civilian defense idea but the public never bought it. The Office of Civilian Defense was subsequently transferred into the Department of Defense. A number of fallout shelter signs have appeared throughout the country, mainly on buildings in the central cities. Some of these fallout shelters have a few supplies of food and water, but for the most part our civil defense program has been nothing more than the process of nailing up signs. It has been quite a boon to sign painters and sign nailers.

The situation had actually become so ludicrous that evacuation plans were developed for many metropolitan areas. These evacuation plans have subsequently been relegated to the trash can. The public was simply unwilling to accept the feasibility or the thinkability of nuclear war. The Office of Civilian Defense still exists, but it does not represent any significant part of the national awareness. Since the very early sixties, public discussions of the consequences, possibilities, and ramifications of nuclear war have been completely absent. The civil defense debate apparently convinced the public that nuclear war was too horrible to contemplate.

The civilian defense debate had hardly ended when the Cuban missile crisis occurred. The American public was dramatically confronted with the real possibility of a nuclear war. One would have thought that this incident might have caused the public to change its mind with respect to civilian defense. But there was not even a resurfacing of the debate, let alone a demand for fallout shelters.

No public debate on nuclear weapons spending

Nevertheless, during the 1960s, nuclear weapons programs prospered. In the absence of any public debate, this occurred with essentially a rubber-stamp approval by the U.S. Congress. In a way, the budget requests of the AEC and the Department of Defense for nuclear weapon development have been approved by the Congress with about as much public awareness as the annual request for the budget for tea tasters. Much like the public-at-large, the Congress probably preferred not to think about the unthinkable. It would seem that after the rather scary public debates on civilian defense the public and the Congress had decided to return to the concept of massive retaliation and that the budgets requested by the AEC and the Department of Defense were more or less approved on the assumption that this was the nuclear policy of the U.S. Government.

While there has been an absence of debate concerning civilian defense and the effects of nuclear war since the early 1960s, studies of this nature have, nevertheless, been conducted since that time within the closed-circuit confines of the DOD-AEC complex. It is somewhat difficult to believe that substantial sums of money have been spent over the past several years to count the dead, but a large variety of different nuclear attacks upon the United States have been "war-gamed" to estimate the devastation and death that would result.

The think tanks fight nuclear wars on paper

In the early 1960s, Mr. Kahn and others were discussing nuclear wars in which some 1,500 megatons were dropped on the United States. At that time the outcome of these wars resulted in the death of some 90% of the population of the United States. In the latest reports from the RAND Corporation, Stanford Research Institute, Institute for Defense Analysis, and other think tanks, the wars which are being gamed today consider up to 12,000 megatons being delivered. Strangely enough, although the number of megatons that are war-gamed have increased some 10-fold, in some of these games more than 50% of the population of the United States survives a nuclear attack. Thus, in the absence of the public debate concerning the effects of nuclear war, the nuclear weapons have become more sophisticated, the number of megatons that would be delivered upon the United States has increased something approaching 10-fold, and yet the people who count the dead would suggest, today, that more people would survive the vastly greater war.

One of the reasons why these paper studies conducted by the think tanks for the DOD-AEC complex would suggest that nuclear weapons are less effective today than they were in 1960 is that the war games which they are playing now have different rules. One change in the "rules" results in individuals surviving in fallout shelters. Another major change has been that in order to make nuclear war thinkable, the military strategists have decided that the other nation, the one which will attack the United States, will deliver its first salvo of missiles against our strategic deterrent. As a consequence, the targeting of these devices concentrates them in different parts of the country more than the weapons which were involved in the wars being programmed or played in the early 1960s.

However, if we seriously examine these paper studies, we discover something which should come as no surprise to anyone: the situation is still the same as it was in the early 1960s. With or without these paper wars, most of the population in the United States would be killed as a result of a nuclear war, and quite possibly in the aftermath, the living would envy the dead.

It should be remembered that 50% survival figures are nothing new. During the civil defense debate, Herman Kahn and others also indicated that if we spent this or that number of billions of dollars, we could save 50% or more of our population. Nevertheless, recognizing that hundreds of millions of human beings would be killed or maimed, that one of the legacies of the nuclear war would be genetic change which would produce severe consequences for more than a thousand years, and that a society emerging from its fallout shelters into a ravaged landscape could not be the same society that entered the shelters, the American public summarily rejected civil defense.

They try to make war feasible and safe

What seems incredible is that the paper studies of these closed-circuit groups who have been studying and developing the emerging new nuclear strategy of this country indicate that for some time they have been convinced that nuclear war is a feasible human endeavor. As an example of the extent to which they have gone in order to plan and to count the dead and the living, we can cite one report which deals with the individuals who have gone for shelter in a 2-story building.

This building has a basement and two floors above the basement. Now, the fallout from the nuclear war will fall on the roof and the ground surrounding the building. The nature of the physics in this situation is that the basement offers the greatest reduction in dosage to the individual, and as one moves up to the first floor, that is still preferred to the second floor. Therefore, in order to keep these people alive, the planners devised the ingenious plan of moving the people who are in the basement stepwise to the second floor where they stay for a period of time, and then subsequently rotating them back down to the basement. In this way the planners were able to average the dosage to the individuals in the building to a point where they figured that all of the individuals would survive.

Another example of the extreme to which such a closed-circuit group has contemplated the post-war situation, let us consider a few quotations from a report prepared by the RAND Corporation.

> ... Furthermore, it would be impossible to limit preferential treatment to labor-force members alone, for the working members of society would insist on transferring some part of their personal advantages to members of their families who were not directly contributing to output.
>
> Policymakers would presumably have to draw the line somewhere, however, in making such concessions, and those most likely to suffer are people with little or no productive potential: old people, chronic invalids, and the insane. Old people suffer the special disadvantage of being easily identified as a group and, therefore, subject to categorical treatment. . . . In this sense at least, a community under stress would be better off without its old and feeble members. . . .
>
> The easiest way to implement a morally repugnant but socially beneficial policy is by inaction. Under stress, the managers of postattack society would most likely resolve their problem by failing to make any special provision for the special needs of the elderly, the insane, and the chronically ill. Instead of Medicare for persons over 65, for example, we might have Medicare for persons under 15. Instead of pensions, we might have family allowances. To be sure, the Government would not be able—nor would it be likely to try—to prevent the relatives and friends of old people from helping them; but overall, the share of the elderly in the national product would certainly drop.[70]

These quotations would suggest that if you are going to think about nuclear wars, you have to be as cold-blooded as a snake. This is why most Americans refuse to consider nuclear war as a feasible human activity. Nevertheless, since the early 1960s, the Department of Defense and the Atomic Energy Commission have sponsored studies that considered, in meticulous detail, the effects of nuclear war and the post-war recovery era. The nature of these studies can at best be described as a valiant attempt to make nuclear war a feasible human endeavor. The tragic thing about these studies is that they undoubtedly form part of the thought processes of many members of the Pentagon who are involved in developing our present-day nuclear strategy.

On the other hand, a number of individuals within the Pentagon and the Atomic Energy Commission have made decisions concerning the development of various nuclear weapon systems without *any* real consideration of the biological effects of nuclear war. It would seem that they left decisions concerning the real need for these systems to other individuals. As an example of this, we would simply offer the directors of the Lawrence Radiation Laboratory at Livermore. This laboratory is the major weapons laboratory in the nation. As a matter of fact, there is only one other weapons development laboratory, and that is Los Alamos laboratory, but Livermore carries the lion's share of the weapons development effort. That's the business of the Lawrence Radiation Laboratory at Livermore—the development and testing of nuclear weapons.

Nuclear war planning ignores biological effect

But strangely enough, in December of 1968, Tamplin was asked to brief the directors of the laboratory on the effects of nuclear war. At this particular time the directors of the laboratory had come out strongly in support of the new military systems, Multi-Reentry Vehicle (MIRV) and Anti-Ballistic Missile (ABM). As a consequence of the discussion subsequent to that briefing, it became apparent that while the directors of the Livermore laboratory were strongly supporting the development of these new weapon systems, they had very little concept of the devastating nature of our own, or the Russians', existing nuclear capability. Moreover, Dr. Michael May in setting the guidelines for the briefing indicated that he was only concerned with the immediate survival of the people. He was not concerned with the long-term biological effects or the genetic consequences of the radiation.

It is a real tragedy of the modern day that the directors of the Livermore laboratory, with their extremely parochial attitudes, should have so much influence in the policies of this nation that relate to all mankind for all time.

Contrast the directors of the Lawrence Radiation Laboratory with Dr. Freeman Dyson, a physicist from the Princeton Institute for Advanced Studies. In the April 1969 issue of the *Bulletin of the Atomic Scientists* Professor E. J. Sternglass of the University of Pittsburgh published a report, already discussed, in which he suggested that the fallout from nuclear weapons tests was responsible for 400,000 infant deaths in the United States. In that same issue of the *Bulletin* Dr. Dyson had published an article that represented a very strong support for the proposed ABM system.[71] In the June issue of the *Bulletin* Dr. Dyson submitted a small communication in which he indicated that his arguments favorable to the ABM system were insignificant in light of the article by Sternglass. He indicated that the Sternglass article was a substantial argument against the ABM system.[72] Confronted with the potential biological effects of radiation, this physicist modified his stand on the ABM system.

Spending billions to protect missiles rather than people

Thus, although there has since the early 1960s been an absence of public debate concerning the effects of nuclear war, the Department of Defense and the AEC have been proceeding since that time with the concept that nuclear war is a feasible human activity. The development of the ABM concept, the MIRV, and the whole present nuclear strategy of the military has occurred since that time, aided and abetted by macabre studies conducted by members of this closed-circuit community. The fate of all mankind is being decided by a fairly small group of individuals who have performed their paper studies in the complete absence of public scrutiny.

Consider the absurd situation in which we now find ourselves. The Pentagon was not allowed to plan to fight a nuclear war through a multibillion dollar program aimed at protecting the population with a massive civilian defense effort. But the members of the Pentagon proved to be quite resourceful and today we find them engaged in a multibillion dollar program to protect a bunch of missiles.

Many members of the Congress and the public must have been amazed when the AEC and the Department of Defense proposed the Anti-Ballistic Missile Program. It must have come as a shock because these members of the public and the Congress suddenly realized that the Pentagon and the AEC had been proceeding since the very early 1960s with the concept of the thinkability of nuclear war. By proposing the ABM system they have put the fat back into the fire. Many senators, realizing what was taking place, made a determined attempt to defeat the ABM proposal. Unfortunately, they lost by one vote.

We were that close to ending the arms race—just one more senator. We could now be embarked on one of the noblest social experiments of all times—an experiment designed to end the terrible threat of a nuclear holocaust and to bring about world peace. Let us hope that before too many more elections pass, this nation will have been able to elect that one more senator. Herein must lie the greatness of America, a nation willing to take the risk necessary for peace on earth.

11 Moral and social responsibility of science and scientists

The radiation-hazards controversy is of monumental importance to man and his ability to survive on this planet. It may, therefore, surprise the reader when we say that this issue is totally dwarfed by a vastly larger problem, of which the radiation hazards issue is but a reflection. That larger problem goes to the heart of the matter, not only survival on this planet, but the quality of existence on earth. We refer to the truly significant crisis involving the social and moral responsibility of scientists and technologists for their actions. The radiation hazards controversy is now in the open, where no amount of subterfuge will again hide it. It will be resolved. The personal fate of the scientists involved in this controversy is a trivial matter in any larger sense.

But the general role of science and technology in society is now in sharper focus than ever. The radiation problem is simply illustrative of the dilemma we reach when scientists and technologists fail in their responsibility to society. It may seem a trivial truism that science and technology should serve society. But has it in the past? And does it now? If the answers to these questions were an unequivocal "Yes," there would be no environmental crisis occupying a prime position in our thoughts today. And why is the answer truly "No"? This we must examine in its several important facets, interlocking in character, and all of supreme importance for a viable, free society.

The phenomenal accomplishments of science and technology over more than a century have certainly revolutionized the way we live. That some of the changes are clearly beneficial to man is undeniable, although only shock and dismay should properly be our reaction to the unevenness with which these benefits have been distributed. But surely advances in agricultural methods, in medicine, in production of industrial items can all be regarded as manifestations of enormous talents of men. Admiration of such talent is inescapable.

Conversion of such admiration into a cult of worship of science and technology is an unmitigated disaster for humanity—and by no means solely in an ecological sense. For the cult of worship of science and technology generates two important idols: (1) Self-worship of the scientists and technologists, and (2) Worship of science and technology as gods in themselves.

The effects of establishment of these two idols are pervasive and destructive in some ways that all now realize—simply by looking around at the neglect we bestow so magnanimously upon so large a segment of humanity.

Science fails to meet societal needs

Science and technology have failed in the task of solution of key problems. We see demonstrable poverty for 50% of our people in the U.S., no matter what the official statistics. We see disenfranchisement of a major fraction of our citizens, no matter what the voter registration ostensibly shows. We see alienation of the bulk of those to whom the future should belong—if indeed even the future is left for them. These effects are self-evident. But the subtle effects are even worse, precisely because of their subtlety which allows for the development of a saintly halo around scientists, and this virtually insures that the worst possible errors of science and technology are not only permitted, they are *encouraged* and made ever more pervasive. The errors become pillars of the culture itself, above and beyond the non-contribution to meeting of key societal needs.

It can hardly be stated that many human beings of intellect devote their lives to evil motives or goals. And it is totally counter-productive to use such a premise in the endeavor to understand the ultimate deleterious effects of science plus technology *and* their practitioners. Let us eliminate all consideration of bad motives, of impugning the motives of any single scientist or technologist. Indeed, let us credit every decision, every action with the best of intentions. Does this give us reason for confidence, for reassurance that the interests of society will be paramount?—that societal needs will be met? Hardly, for the cult of self and entity worship develops an inevitable dynamic of its own, answerable to no outside influence, and characterized by a relentless pursuit of its own ends. And all the while a self-righteousness proclaims repeatedly that everything is being done to create an utopian dream for man, if he would but understand and appreciate the devoted attention to his welfare.

Why are we surprised that a central feature of a cult of worship of science and technology should be the concept that more and more of the same science and technology are obviously needed, desirable—indeed the *sine qua non* of a better world? For are not the works remarkable, breathtaking, almost fictional? Indeed they are—and it is but a short step to the assumption that what science and technology cannot do isn't worth doing— a short step to the view that no matter what uses are made of science and technology, no matter what dire effects may thereby be created, science and technology will surely provide an answer. Omnipotence is the essence of the idol. And so it is that for every problem created by such a cult in its self-aggrandizement, it is assumed there must be a solution available—available through the very agencies and forces that produced the devastating problems in the first place.

It is not at all unexpected that science and technology in the service of industry and government, having led us to the brink of so many disasters, should be ready to proclaim proudly that in the interests of the general welfare these same stalwarts are prepared to provide us with their services in providing a solution. There is not a single instance (or few at best) where this sequence is not descriptive of the real-world story.

For transportation-in-the-ultimate we were blessed by the

automobile with its infernal combustion engine, for every man his own personal dream. And a dream it must be, sexy, in the color of his choice, or two, and replete with devices about as related to efficient transportation as a head cold is to a happy day in life. And now that we have come to realize that it is either us or the automobile, we are promised solutions are just around the corner, for technology will develop fuel additives or exhaust devices that will assuredly bring us back to that simple ideal of breathable air.

Development of nuclear weapons to insure 'peace'

For defense, the nuclear weapon is the greatest bargain in bang-for-a-buck yet conceived by man, and who is to deny that big bang potential will assuredly drive off would-be enemies? Shortly after World War II the pronouncements were pontifically made that with this advance, war was no longer thinkable. Science and technology had wrought the everlasting peace so cherished and so evanescent for so long. But soon the clouds gathered as the realization dawned that the science-technology cult existed in more than one geographic area of the globe. Two, it was found, can play at this game and the fear returned, only larger, that war not only was possible, it held potentialities never before realizable.

Quickly to the rescue, science-technology achieved the remarkable answer that more and more nuclear weapons could be built, by dedicated effort; bigger, too, a thousand-fold more destructive, again guaranteeing a secure peace—until too many powers joined this elite set of guarantors. Then, deliver them faster—the miracle of the missile. But "they," too, have missiles. Then create an anti-missile—no problem. Here assuredly was the ultimate in humane considerations.

The nuclear weaponeers foresaw a new golden era—no longer man against man, but machine against machine. Brilliant beyond belief! For MIRV, the answer to anti-missiles, we can assuredly count on some form of technology-born anti-MIRV. Those who question whether this whole sequence has increased *anyone's* defense or security simply don't understand what science and technology have to offer. This is a solution, we are assured, if we will but support the noble science-technology effort to the almost total exclusion of every worthwhile human endeavor.

If the weather is bad, science-technology will assuredly learn to modify it. And if such modifications lead to disastrous ecological changes, leave it to devoted research and development to find a solution.

If science and technology have increased the earth's bounty in crops, an agricultural revolution, we know we should have counted on similar brilliance to develop the pervasive defoliants and herbicides to guarantee that crops at all will no longer be possible in those areas of the world we so avidly decide to civilize.

The SST a 'monstrous abortion'

Where Leonardo da Vinci failed to emulate the bird, modern science-technology has given man those long-sought wings; and the assurance that he can fly ever faster, in ever larger boxcars. Why? Because it must be good for something or someone, he is told. The economy will be stimulated, jobs performing unnecessary tasks will be available, and the national pride will be assuaged. So now man enters the era of the super-sonic transport, a device so unacceptable as to create an unbearable noise pollution. But have no fear, what havoc science-technology has wrought, by determined effort and abundant infusion of tax-dollars from those who will never benefit from such monstrous abortions as the SST, scientific research will assuredly find solutions to extricate us.

The undying confidence in technology's ability to undo the multitude of rapacious insults to man and his environment so freely and efficiently created is beyond belief, so arrogant has it steadily become. In the field of atomic energy some ultimate examples are available. Speaking glowingly of the future of production of nuclear electric power, Dr. Arthur Upton, a pathologist at Oak Ridge National Laboratory, tells us that as a

Moral and social responsibility of science and scientists 205

by-product of nuclear reactors for production of such power, we shall have billions of curies of accumulated radioactivity to dispose of by the year 2000.[73] At the outset let us state we have a very high regard for Dr. Upton *and* his scientific work. We question whether he has adequately considered implications of his statements.

Dr. Upton tells us that information on disposal of radioactive wastes is limited. Hence, he states, the problem of disposing of such wastes will assume increasing proportions with the development of nuclear power in the next several decades, since this will require the disposal of many more tons of radioactive waste than are now being disposed of annually. And listen to Dr. Upton carefully as he tells us that "The methods envisaged for disposal call for permanent storage of part of the waste on land or underground and discharge of the remainder into the atmosphere and into rivers and oceans."[74]

The rape of our atmosphere, our land, our rivers, and oceans by radioactive pollution forever is not even questioned at all by Dr. Upton. Dr. Upton does not ask himself or us whether we are rational human beings to commit this obscene act against man and the earth. He simply accepts it, for who shall question the decision to go ahead with nuclear power generation accompanied by this outrageous radioactive waste disposal plan? Science-technology has so ordained! The cult has spoken! We come to understand Dr. Upton's view of solutions in his following statement:

> Although present methods for disposal of radioactive wastes appear to be satisfactory to meet existing needs, if closely controlled, the effects of discharging greatly increased amounts of radioactivity into the environment cannot be predicted without more information on the behavior of fission products in the biosphere and about the effects of fission products on man and his ecosystem.[75]

Dr. Upton accepts the unacceptable disposal plan and hopes human beings and the ecosystems supporting them will *somehow* muddle through if we get more information. Not a single suggestion that maybe we should use our brains to *avoid* this calamity. And in the best tradition of the science-technology cult of self-worship, he goes on as follows:

> The disposal of increased quantities of radioactive waste will thus call for additional research, as well as further coordination and refinement in environmental monitoring.
>
> These questions undeniably will constitute a growing challenge; however, the comparative safety of the atomic energy industry to date encourages hope that the challenge can be met successfully if the necessary planning and research receive the attention and support they deserve.[76]

The unbelieving reader should be reminded that Dr. Upton has been responsible for *excellent* scientific researches on radiation induction of cancer. And his motives are undoubtedly the best. Such are the pernicious and pervasive effects of the science-technology cult upon the practitioners that statements such as his are not only possible, they are commonplace.

Research, we see, will be necessary to learn, to the third decimal place, an answer we should long ago have rejected as abhorrent. Indeed, we can learn this marvelous answer if we but devote ourselves to it. How such radioactive poisons become thoroughly spread into the biosphere, including man, we shall study in detail, even though we know only too well (in large measure through Dr. Upton's efforts) the deadly effects that necessarily will eventuate. All we must do is support research generously; for if we do, assures Dr. Upton, we can "hope" that science-technology will yet bring us out of the greatest morass it plans to create. And thus we may arrive at the miraculous triumph of science-developed anti-pollution in the valiant attempt to keep pace with atomic technology's murderously efficient rate of creation of irreversible pollution.

Do not prevent the pollution itself—that might offend the almightly technology itself. Instead, we are asked to *study* the pollution, where it goes, and how it kills life on earth. By dint of dollars and devotion, we shall overcome!

Dr. Upton is not alone among those who accept nuclear

pollution without question. We repeatedly hear from atomic energy disciples, *after* the pollution of rivers, streams, and lakes, that they have thereby created "living laboratories" for the study of radioactivity's effects upon living things. How extremely thoughtful and generous!

And this is the problem of the irresponsible arrogance of science-technology.

Once heralded as a possible deliverer of mankind from the bondage of want from hunger, disease and lack of shelter, and as a harbinger of the good life, it has long ceased to prove itself operating in society's behalf. It operates by its own rules, with allegiance to no human values. What science and technology has done is to make itself the goal, the objective—more, bigger, better. It can do no wrong. And if, by chance(?) it has, assuredly its unequalled brilliant efforts, always *thoroughly* supported by the labors and taxes of others, can extricate us from disaster once more.

Neutrality not always the same as responsibility

Such is the dynamic of this cult of worship of science-technology. And it sings its self-praises of objectivity, adherence to truth, and its ethical *neutrality* in the matters of men. It knows no international boundaries, and can thereby reveal to men of all countries the laws of science by which they can retreat from responsible behavior to themselves and to others. And such is the self-aggrandizing power of the dynamic that it brooks no questioning, and it tolerates none, while it speaks of its role in preserving and extending "civilization."

And the second danger—that the scientists and technologists come to worship themselves as idols—what effects does this have upon their social responsibility? This is worthy of examination. Not only does science-technology imagine itself omnipotent; so, too, dream its practitioners. Dreams of omnipotence inevitably breed a second evil, self-righteousness. And under this cloak can unfold a behavior of men governed by a dynamic of irresponsibility par excellence.

Professor George Wald* commented, in answer to those who reassuringly said nuclear weapons were just "a fact of life," that "these are not facts of life; they are the facts of death." How correct he is! It is such acceptance, without a profound soul-search, that is a hallmark of the perversion of, or failure to understand, human values and obligations, and which leads to the deep immorality of the practitioners of science and technology. Not a planned immorality; indeed, self-righteousness leads them to express their horror that anyone would be so uncouth as to even raise questions of their motives. We are not questioning their motives as evil; rather, we are questioning the very *absence* of motivation to truly moral behavior, a behavior that implies a deep reflection upon the implications of one's actions.

The arrogant self-esteem of scientists

Atomic energy, peaceful and military, provides gross and extremely important examples of what happens to scientist-technologists. We have had the opportunity to live through, at close contact, experience with both examples. One of the characteristics of self-worship as an idol is, of course, the developing feeling of omnipotence. And with this comes the arrogance that suggests no self-examination is required. With this arrogance there is never a requirement to ask, "Are we asking the right questions?" "Are we beginning at the beginning of the story?" or, in our arrogant self-esteem, "Are we accepting premises without examination and then winding down an inevitable path to unseen, but terribly real, disaster in human values?"

The real atomic era, of course, begins in 1945 with the explosion of atomic bombs over Hiroshima and Nagasaki. Man had, at last, unlocked nature's secrets to enormous stores of available energy. It is not an unreasonable step to the statement that these vast energy resources will now be applied to the betterment of life on earth. Yet along with these stores of energy come

---

*Dr. George Wald is Professor of Biology, Harvard University, and Nobel Laureate in Medicine, 1967.

by-products capable of eradicating essentially every form of life on earth.

Were there a lesser degree of arrogant assumption of omnipotence, scientists and technologists *might* have asked the question, "Is there a way to utilize these riches of energy *truly* for man's benefit, or may these by-products, these radioactive poisons, thereby create 'a hell on earth' for man and other living creatures?" But this question is not asked, and with righteous self-assurance we press forward toward widespread use of atomic technology. Our experts in the biological and medical sciences will assuredly, through dedication and well-financed efforts, enable us to cope with any and all problems which arise. And what is the first problem—and the first compromise? Obviously, how careful must we be in keeping the by-products from intersection with the biosphere—man, animals, plants? When we say this is the *first* problem, we mean precisely this—it *is* and *should be* the first problem. But it hasn't been.

For substances of fantastic potential for injury to life, understanding of such potential in all its short and long term ramifications would seem the first step. But such understanding comes only with careful, patient study, reflection, and further study. With some humility, one also realizes that beyond our accumulated knowledge at any particular time, there may be a vast sea of ignorance. Does this lead, or has this led, to an appropriate degree of caution on the part of biological scientists involved in atomic technology? We consider the answer to be a resounding, "No!" And we would hasten to add that, earlier in our own careers, we were certainly as culpable as any other biological scientist in this regard; we surely shared, unwittingly, the arrogance of assumed omnipotence of science and scientists.

"Progress," reasoned the biological scientists, "must be made. Progress has brought us every good thing in life. And, further, all life has hazards; we simply learn to cope with them. and by our tireless and devoted efforts, we will cope with the hazards of technology's poisonous by-products. But such tireless efforts will come later. Today we must provide atomic technology with a set of rules, or guidelines, concerning how much exposure we shall allow human beings to receive from radioactive by-products as atomic technology advances."

Economy more important than health

How much is acceptable? Few person are so callous as to consider an obviously fatal exposure as acceptable. Indeed, we can be assured that such disrespect for life characterizes none of the biological scientists involved. A dilemma is perceived; our knowledge is scant; our ignorance is large, but a set of ground rules are being clamored for by the atomic developers. Can *we* tell them to await acquisition of the knowledge? Can *we* tell them to conduct their technology with no exposure of humans to radioactive by-products? How will we answer their charge that such a requirement might "price atomic technology out of business"? "We must," say the omnipotent biological scientists, "be realistic. We must operate in the real world, cognizant of the facts of life (or could it be 'death'?)."

And so, from a melange of knowledge plus ignorance, a set of radiation exposure standards are arrived at, and proclaimed as "acceptable." Of course, those who promulgate such standards are sincere. Of course, they work hard, and would be horrified to feel they were harming fellow men. But where is the moral self-questioning, the responsibility for actions that justifies saying, "We must live with the facts of life—and the facts of life are that atomic technology must proceed efficiently and *economically"?* Economically? For whom? For what goals? At what cost in life or the future of life on earth? The word is magic—we simply must proceed *economically* without delving too deeply into the meaning of the word.

Listen to the pronouncement of the International Commission on Radiological Protection as recently as 1958 in justifying the standards they recommended for human exposure to radiation as acceptable: "The Commission believes that this level (5 rems per generation) provides reasonable latitude for the expansion of atomic energy programs in the foreseeable future."[77]

"Reasonable latitude for the expansion of atomic energy programs." What, we must really ask, does "reasonable latitude" mean? Unfortunately it relates to the word "economically"—or expressed otherwise, what won't annoy the enterprising technologists too much with costly constraints to protect human life. Is there any other interpretation possible? But note the qualifications as the ICRP admits its insecurity about the size of the risk factor:

> It should be emphasized that the limit may not in fact represent a proper balance between possible harm and probable benefit, because of the uncertainty in assessing the risks and benefits that would justify the exposure.[78]

We have here an admission that a decision has been made to allow injury to man's greatest treasure—his genetic inheritance—without knowledge of the magnitude of such injury. For what do we accept this unknown risk? For "probable benefits." To whom?

How shall we view the common justification for this choice of "permissible" genetic harm so often expounded by biological scientists in the atomic field? "Man," they tell us, "has managed to get where he is receiving approximately 3 to 5 rems per generation from *natural* radiation sources; surely, *adding* an exposure equal in size will not be serious." When will we know if this statement seals the irreversible doom of the human species? It's too late by the time the species disappears. Where is the evidence, beyond pious self-assurance, that this does not constitute a prescription for genocide?

Loss of responsibility to mankind

One cannot criticize the sea of ignorance, the gaps in our knowledge. But what shall one say about scientists setting a policy permitting radiation of hundreds of millions of humans, an experience never before encountered, with a justification of "providing reasonable latitude for the expansion of atomic energy programs?" The answer is simple. These scientists had learned the "facts of life" all too well. But they were and are of the highest motivation by *usual* standards. Are usual standards good enough? We must answer with an unqualified, "No!" For the "facts of life" will indeed be the facts of death if scientists fail to realize the meaning of their overt and covert subscription to their own omnipotence.

Our recent experience with human frailty in Nazi Germany should certainly have taught us the lessons we ostensibly taught the world at Nuremberg. A higher law, a higher morality, we told the world, exists; it must be adhered to by men, no matter what orders or directions are given them by governmental or other superiors. They must, we said then, obey that higher law—responsibility to their fellow men and humanity's larger goals. Has that lesson been lost? Lost, especially, upon scientists and technologists? We believe it has. And all the more cruelly so because the public, astounded by the magnificent accomplishments of science and scientists, has been willing to place its faith in such men, never realizing that the higher law might be the "orderly and economic" development of a particular technology.

We must examine this further if we are truly to appreciate the ultimate, and highly significant, ramifications. The statement above was made by the ICRP in 1958, and shortly thereafter the U.S. Federal Radiation Council adopted these suggestions for radiation standards to apply to "peaceful" development of atomic energy. And since then the evaluation of our "sea of ignorance" biologically and medically has been fantastic!

After the standards were set, a whole new field of medical importance developed—almost entirely in the period beyond 1960—the field of human cytogenetics, or study of human chromosomes. This body of medical knowledge has accumulated rapidly during the 1960-70 decade, and while impressive in its accomplishments, it is vitally more impressive by the vistas of as yet undiscovered facts of importance for human health and disease. In these chromosomes are carried *all* the information (in genes) which instruct a fertilized ovum to become a human being, and these chromosomes are visibly injured by exceedingly small amounts of radiation. There are 46 such chromosomes in every human cell (23 from the mother; 23 from the father). At this writing, no one knows: (a) what crucial biological genetic information is carried by each of the 46 chromosomes; or (b) how sensitive each chromosome is to breakage by ionizing radiation; or (c) what the implications are of small chromosome breaks, and hence losses of chromosome pieces, by radiation in sperm or oval precursor cells. What will this do to fertility? What will this mean for diseases and defects in future generations? How many such broken chromosomes are consistent with producing new generations of humans at all?

Radiation standards are based on old data

So, perhaps *the* most important aspect of human medicine is now in its groping infancy. Yet, our standards for radiation exposure of these delicate chromosomal structures were set *10 years before* this explosion of now-accumulating evidence. And our scientists, with characteristic omnipotent confidence, tell us our radiation standards are "safe"—that careful, conscientious scientists toiled to establish these standards, and *therefore* they must be safe. We do not question the carefulness of the scientists. We do not question that they toiled diligently. Nor do we raise a single question concerning their conscientious application of inadequate knowledge then available to the task before them.

But we know they were working in a sea of ignorance—no fault of theirs; the knowledge had not yet arrived. And we are truly appalled, not by their diligence, but by the arrogance of omnipotent self-idolatry which is reflected by their lack of understanding of the potentially massive hazard at which they were, unknowingly, placing the human species—to guarantee "reasonable latitude for expansion of atomic energy programs." What conceivable good expanded atomic energy programs will do for a world without humans escapes our understanding.

Bygones should, of course, be bygones. Past errors should be considered only insofar as they help us avoid a second round of errors. But this is not what is happening. The errors are simply being compounded and extended.

Mindful of this new body of evidence concerning genetic and chromosomal disorders, we wrote, in presenting our original evidence in this volume concerning cancer + leukemia hazard from radiation, the following:

> ... And we must add to these estimates the comment that we have used only the *hard data* in hand based upon cancer and leukemia induced in humans by radiation. We have said *nothing* of the additional possible burden of loss of life and misery from genetic disorders in future generations, fetal deaths, and neo-natal deaths.

And further:

> Any conclusion we draw concerning the hazard of the current radiation guidelines can only be amplified and buttressed by consideration of human misery associated with genetic defects, fetal deaths, and neo-natal deaths. The case against perpetuation of the existing FRC guidelines is overwhelmingly strong just on the basis of the cancer-leukemia risk, without even considering the potentially *much larger* problem of effects upon future generations.

We did, indeed, appreciate the "much larger problem" of genetic disorders, and possible chromosomal disorders, but because we then knew our ignorance prevented even an estimate of size of effect, we refrained, in the face of ignorance, from saying anything about the size of this calamitous possibility.

And were we alone in our concern about the unknowns in the picture of hazards to humans from radiation in 1969—almost a decade *after* the "safe" standards were set? Let us look at the words of one of America's most respected students of such matters, Professor Brian MacMahon of Harvard University, writing in 1969:

> While a great deal more is known now than was known 20 years ago, it must be admitted that we still do not have most of the data that would be required for an informed judgment on the maximum limits of exposure advisable for individuals or populations.[79]

Moral and social responsibility of science and scientists

The reader will realize, from what has been stated in the foregoing, that Professor MacMahon is, of course, completely correct. He realizes very well that new discoveries in genetics, in human chromosomal cytogenetics, are in their infancy. How can we possibly know what radiation dosage is advisable or acceptable if we don't know the magnitude of effects other than to realize they can be many, many times as large as the shocking cancer-leukemia risks.

Dr. Tompkins worries about costs

And how does the moral and social responsibility of some of the standard-setting scientists manifest itself in response to our warnings of the serious defects in our standards? Warnings of the potentially disastrous effects of allowing atomic energy to proceed with such allowable radiation dosage to our people?

Dr. Paul Tompkins is presently the Executive Director of the Federal Radiation Council. How did he respond to the Gofman-Tamplin suggestion of a massive reduction in the amount of radiation the government allows for Americans? Let us quote Dr. Tompkins:

> It (the Gofman-Tamplin recommendations) might well price society out of business. To reduce radiation exposure ten-fold would cost billions; it might even cost more than the Vietnam war. To comply, you'd practically rebuild all nuclear installations and the factories that use any sort of x-ray equipment. We'd have to review radiation exposure from wrist watches, TV set and radium dials. Plus, I'm not completely sure it is now technically possible to monitor down to such a tight level.[80]

We have here the characteristic response: The *economics* of safety. The cost in *dollars* in preventing human suffering. Would it be amiss for citizens to ask their "protectors" to be more concerned with health and safety, and less with dollars and economics?

The confusion and disarray in AEC scientific circles is, however, far more ridiculous than the above. Apparently not knowing or remembering what Dr. Tompkins had said about costing "billions" to reduce radiation exposure 10-fold, Dr. Victor

Bond* and Dr. Theos Thompson, two staunch AEC scientists, not too long after, hastened to assure us that no AEC programs either are going to, or are contemplating giving even a small fraction of the allowable radiation exposure, even out to the year 2000. If we listen to Dr. Bond and Dr. Thompson, the cost of reducing standards 10-fold would not be *one* penny, since AEC programs (according to them) have no possibility of giving anywhere near this exposure to people.[81] How strange, indeed, is Dr. Thompkins' statement that reduced standards would cost "billions." Which of these confusing standards shall the public believe?

Not one penny—or billions of dollars. That's quite a range, it seems to us. But such confusion among AEC and FRC officials is only the beginning. They are contradicting each other concerning delivering *1/10* of the current allowable radiation dosage, while along comes an AEC official who is quoted as saying he doesn't know how soon the *full* current "allowable" limit will be reached by the U.S. population, and adds it "depends upon Plowshare programs, among other things."[82]

Confusion in the AEC

Surely *someone* in the Atomic Energy Commission or the Federal Radiation Council must have some idea of what the radiation exposures are or are expected to be from current practices—better than statements which imply: (a) We'll never even by the year 2000 approach more than a small percent of the current standards; (b) It will cost "billions" to reduce the current standards to 1/10 their current value; and (c) We can't say *when* everyone will reach the dosage allowed by current standards.

Might the public ask for a little more credibility from AEC scientists and spokesmen? What impression might the public get from such confused, contradictory statements concerning social responsibility of AEC spokesmen and scientists?

*Dr. Victor P. Bond is Associate Director, Brookhaven National Laboratory.

The Atomic Energy Commission scientists and supporters wasted little time before leaping to the defense of current radiation standards—those self-same standards that have no scientific foundation whatever and which were decided 10 years ago—before much or most of the revelant evidence was in (and that evidence in many respects just *beginning* to come in as far as the chromosomal and genetic effects). Let us look at the pernicious effects of the arrogant self-adulation of scientists, well-intentioned as it may be.

Congressman Holifield is currently Chairman of the Joint Committee on Atomic Energy—supposedly a "watchdog" committee, but now under the aegis of Holifield and Congressman Hosmer, even more *promotional* that the AEC itself. A group of scientists, apparently worried that Congressman Holifield might be suffering from sleepless nights over his concern that lowered radiation standards could interfere with his favorite AEC promotional schemes, decided to write Mr. Holifield a delightfully soothing and reassuring letter of sympathy and condolence:

> Several reports have appeared suggesting that the authorities responsible for the guidelines for the safe uses of ionizing radiation have been grossly complacent and even in error in setting their current radiation standards. Unfortunately, adequate rebuttal requires a somewhat lengthy and technical reply unsuitable for publication in the press.
> Such material as is necessary is contained in the publications of the Federal Radiation Council (FRC), the National Council on Radiation Protection and Measurement (NCRP) and the International Commission on Radiological Protection (ICRP). These reports show evidence of the great competence of these bodies, and their concern for the public health.[83]

What do the 29 signers of this condolence letter really tell us? First, lest there be any misunderstanding, it would be a grave mistake to impugn the motives of the people who signed this letter, or to impute evil intentions to them. They can, however, be regarded as outstanding examples of the dynamic of science and scientist idolization and self-idolization. The maxim of this dynamic is that science must be intrinsically good and that scientists, no matter what their actions, must somehow be acting in the best interests of the public. If this had indeed been the case over the past half century, we would now be facing no crucial environmental crisis of deepening proportions. But we are, and science-technology will find no place to hide from its responsibility in helping to generate the crisis, no matter what pious phrases its apologists make. Again, no evil intent, but an adhorrent result of a relentless dynamic. We may translate the Sagan letter (above) as follows: "The people who populated the various Commissions and groups who promulgated the various standards, now under total challenge, were sincere, hard-working, dedicated scholars who did the best they could."

Some questions for the signers of the Sagan letter

Why argue with that? Let us assume the motivation was of the highest, the dedication supreme. But let us indeed ask questions about a sense of moral responsibility concerning a non-intended, but nonetheless pervasive, arrogant psychology that characterizes scientists. We must try to understand how 29 scientists justify a set of standards promulgated *ten years before* the most important questions and answers relating to those standards were even known. How do these 29 scientists answer Professor MacMahon's correct statement in 1969 that we don't even now have the necessary information to decide acceptable radiation doses either for individuals or populations? How do these 29 scientists say not only were the obviously unjustified standards acceptable in 1960, but they are even acceptable now, in the face of devastating evidence to the contrary?

What sort of answer, from 29 scientists, to a set of scientific presentations of cancer and leukemia risks 10-20 times higher than previous estimates, is it to retreat from presenting a *single* item of evidence in refutation and to resort to applying the balm of empty reassurance to a troubled Congressman's head?

Eleven thousand scientists joined Linus Pauling in his concern over fallout levels of radiation one-twentieth of those of

Moral and social responsibility of science and scientists

current concern for the peaceful atom. Eleven thousand scientists throughout the world. And since then the evidence has grown even more disturbing.

Yet 29 scientists, almost all within or supported by the AEC, blithely reassure Congress about radiation exposure 20 times as high. And they arrogate to themselves the appellation of a "consensus of informed opinion." This band of 29 scientists apparently considers the vast bulk of the biological community of scientists either uninformed, unconcerned, or incompetent. Or can it be that some dynamic, unappreciated by anyone involved, would account for a group of primarily atomic energy-supported scientists having difficulty seeing the perils of a by-product's hazards of the technology for which they have held such high hopes, no matter how the facts indicate such hopes to have been optimistic and unjustified. Surely the reader can wonder about this problem.

And how shall the 29 scientists explain the justification for the standards they so warmly support as being "consistent with the orderly development of atomic energy?"

Concern over pricing society out of business—Yes!
Concern over orderly development of atomic energy—Yes!
Concern over the welfare of the human species—??

Totalitarianism in science

Defensiveness about one's children is a common human trait. And certainly atomic energy can be regarded as the child of many of the current inbred generation of atomic scientists. The outmoded "safe" radiation standards are the particular child of the bio-medical segment of the atomic energy community of scientists and technologists. Hopelessly trapped by a dynamic which breeds a feeling of omnipotence, and without evidence to support their position, defensiveness is not unexpected—it is simply a human foible once more held up for all to see.

Were the defensiveness simply to serve as a Holifield-balm, one could be unalarmed, amused, and even tolerant. But far deeper issues are at stake—issues that so transcend atomic energy that they must be appreciated by all members of society, for their very life and freedom from totalitarian, authoritarian domination are the real issues.

The potentates of old had, and the dictators of today have, a simple answer for dissident opinion, "Off with their heads." Among so-called civilized men dealing with important scientific issues, technological issues, decisions to go forward industrially, or matters of safeguarding the public health, the vast majority of people in a democracy such as ours would assume that reason must operate. But the day of Salomé is hardly over. And for students of the environmental crisis, in particular those who really would hope to preserve a livable world for humans and other creatures, the realization that totalitarianism is rampant in scientific-technological matters should send a sharp chill through their inner selves. For it is this phenomenon, more than any other, which will block and possibly destroy their most valiant efforts to make a survivable planet of life in all its splendor and beauty.

Is totalitarianism rampant in science-technology? Look at the evidence, using atomic energy as a shining example.

As one of its major charges by the U.S. Congress, the Atomic Energy Commission is obligated to promote the peaceful uses of the atom while giving prime consideration to the health and safety of the public in so doing. The public relations department of AEC, expending unknown but undoubtedly large sums of taxpayer dollars, carefully and repeatedly reassures the public of it noble actions and intentions to give its heart and soul to the task of protecting the public health. The real problem we face is to find that heart, and to examine that will-o'-the-wisp soul—assuming generously either exists.

The protection of the public health means the examination of crucial issues in the development of uses of atomic energy. And the most important issue is the impact of any and all programs and activities of the AEC upon the health and welfare of the present and future generations of humans and the ecosystem of life which supports human existence. For no glorious applications of atomic energy (or any technology) will be a blessing to *non-existent* life! We have explained how a dual charge of seeking out and developing atomic exploitation and protection of the welfare of people leads to a hopeless impasse. Long ago the Atomic Energy Commission should have forthrightly come to the Congress with a frank and honest admission that the two charges were clearly incompatible with each other, and the AEC should have requested it be relieved of one or the other of these two functions. Certainly this would not have been regarded as a manifestation of inadequacy, or of weakness, or lack of ability. The AEC could thereby have gained the confidence and respect of the Congress and the public for a recognition of how important tasks are and are not to be done. But the AEC didn't do this.

In 1963, after a long history of loss of public confidence in its health and safety pronouncements, the AEC felt it had to try to appear to do *something* to restore some semblance of public confidence in its credibility. Even that late, it is clear that the only real answer for the AEC would have properly been to request the Congress to relieve it of all responsibility for any matters relating to public health and safety. A simple admission of total failure in this field would have earned respect and, at least, confidence in its sincerity. No one expects necessarily that the plumber will also be a good pianist; why should they conceivably expect the promoter of sales wares to be a public health protector? But pride and position are, of course, strange and fascinating foibles to observe. And supremely dangerous!

AEC is 18 years late with a bio-medical program

No, the Atomic Energy Commission would not admit its failures to win the public confidence. And this, of course, is why the request was made of Lawrence Radiation Laboratory to set up a bio-medical program, long-range in scope, to evaluate the impact on man and the biosphere of radioactivity release from its various atomic programs. On the face of it, this request was obviously ridiculous. Who, we asked ourselves then, would believe in the sincerity of AEC's concern for public health, when eighteen years after its operation plus expenditure of over a half-billion dollars on "health aspects" of radiation, it is finally deciding to evaluate the impact of radioactivity release upon man. "What," the public could properly be expected to ask, "had the AEC and its already-existing 19 laboratories of biomedical research been doing for 18 years that had failed to learn the impact of radioactivity release?"

Well, the public would hardly have been alone, for all of us who were being asked to undertake this task at Lawrence laboratory were asking each other the very same question, "Why believe any of this?"

"A new leaf is being turned over—let's forget the errors of the past."

"The Lawrence laboratory is a strong, independent laboratory; it stands for truth."

"We want only the truth."

"We at Lawrence will back you no matter how much the AEC tries to suppress the true facts of the hazard of atomic energy programs to man."

"The issue is of supreme national importance—you must undertake the task even though you have little or no confidence in the credibility of the AEC."

Why we "Danieled" into the Lions' Den we shall have an abundant opportunity to reflect upon now that our throats have been thoroughly slashed, and we can repent in leisure what we accepted to do in haste. But, while we can be criticized as naive in thinking the new leaf might be any less disease-ridden than the old leaf, or that AEC wanted any part of the truth, even we did not truly realize the gruesome nature of the Salomé-E-C demands. Nor did we really think our scientific colleagues in the directorate of the Lawrence Radiation Laboratory would melt away like ice cubes in a hot summer sun. That, we felt, would certainly not happen. Optimism dies hard.

And, somehow blissfully ignorant of reality, we still had confidence in the integrity of democratic ideals. While scientific dissent from promoters' ambitions might be unpopular, nevertheless the realization that dissent and criticism in public health matters are vital would guarantee a hearing of the facts—or so we hoped, erroneously.

To those who think the environmental struggle for survival of life on this planet is going to be a proper tea-party, we say, "Think again."

To those who think reason will prevail, we say, "That's a fond dream."

To those who say repression of truth, reprisal against those who try to present the truth, and disdain for the right of the public to hear the evidence, are products of Nazi Germany or other equally despicable dictatorial societies, we say, "Look close to home." It *is* happening here—and in the most shocking of all areas—that of the guardianship of the health and welfare of life on earth!

A reward of derision and slander

You are incredulous? Disbelieving? Then examine the evidence yourself. Look not for evility in men; rather, understand the arrogance and omnipotence fantasies of science, technology and their scientists and technologists. And realize that this is a self-moving force inducing men to respond as pawns, unknowingly, unwittingly, and believing unto themselves they are performing a worthy task for their fellow men.

Tamplin and Gofman presented evidence before a highly respected scientific body that our allowable radiation exposures for the population are grossly unsafe, and could lead to a massive public health disaster.

The AEC response: Derision, denial, slander—but *no* evidence in refutation. Having asked for the study to be done, the AEC ridicules the result of the study since it wasn't the hoped-for result. Science? Responsibility to the public?

Dr. Michael May, director of the Lawrence laboratory, requested of Gofman that *before* he or Tamplin present any further evidence, the AEC be given the opportunity to see it in advance.

He assured them this was no effort at suppression—*he* would prevent any such suppression. But the AEC needed to know in advance so they could be prepared. This *seemed* altogether reasonable. Gofman and Tamplin agreed.

Two short weeks later Tamplin submitted a manuscript on "Nuclear Power and the Public Safety" to be presented several weeks later at the American Association for the Advancement of Science Annual Meeting. Dr. Roger Batzel, associate director of Lawrence laboratory, was asked to forward the manuscript to AEC so they would have plenty of advance knowledge. What happened? The Batzel-censored manuscript was returned to Tamplin with little left in it but the prepositions and conjunctions. (We have this original censored manuscript in our possession, and should probably donate it to the Smithsonian as a future relic of barbarianism in human endeavors—as an exemplary antithesis to free scientific inquiry.)

And Dr. Batzel informed Tamplin that to present the original, uncensored manuscript, he would have to get his own personal typist, typewriter, paper and travel funds to attend this scientific meeting to which he had been specifically *invited* because of his expertise.

Lawrence lab knuckles down to censorship

Gofman reminded Dr. May of his promise of "no suppression" and indicated that Dr. Batzel's censorship of scientific truth meant the end of Lawrence Radiation Laboratory having any right to masquerade as a scientific laboratory. Dr. May replied that Gofman shouldn't feel badly about this little bit of censorship since after all he wasn't doing to Gofman and Tamplin what the AEC had suggested be done. We never asked precisely what the hydraheaded monster did want him to do with us.

Finally, because Gofman threatened to expose the laboratory censorship and suppression at the open meeting of the American Association for the Advancement of Science, the LRL directors backed off, and Tamplin was permitted to present his paper, with modifications greater than he had intended. Incidentally, the meeting was held in Boston, December 28, 1969. In retrospect, it would have been far better to announce the censorship and absence of scientific freedom of Lawrence Radiation Laboratory then and there.

From that point on, the repression and reprisal mounted rapidly. Tamplin had successfully conducted his research efforts with a staff of 12 people. Eight of these staff members were summarily removed from Tamplin's group. The reason, "budget cuts." The laboratory had suffered a 10% budget cut; Tamplin suffered a 67% reduction in staff. *Budget* cuts?

AEC finances trips only if speakers are favorable

Next, Tamplin and Gofman were asked by numerous citizen groups in communities to come to speak concerning the hazard aspect of radiation from atomic energy programs proposed for their respective communities. The AEC lavishly provided speakers to such community meetings to extoll the virtues of its atomic programs and to deny any hazards from radiation—all expenses paid, courtesy of the American taxpayer. But who paid for Gofman and Tamplin to present the hazard aspect of the question?—the citizens *out of their personal meager resources* paid. This, according to the AEC, represents providing the American public with an open forum for consideration of all aspects of the radiation hazards question.

If, as an American taxpayer, concerned about your health and that of your children and further descendants, you are incensed at this concept of "open forum," save some of your anger, for the full truth is even more devastating. The best is yet to come.

When Pennsylvania citizens were concerned that a forthcoming proposed Plowshare nuclear explosive project for their state might be a hazard to their health, the Lawrence Radiation Laboratory stood ready to reassure the Pennsylvanians.

Dr. Bernard Shore, director of the Lawrence Radiation Laboratory Bio-Medical Division, and two other bio-medical scientists, were paid their salaries and all expenses to go to

Pennsylvania to reassure the citizens. A panel discussion was held at Pennsylvania State University on April 17, 1968. This, of course, the Lawrence Radiation Laboratory regarded as a proper work assignment. Of course, we must add that Lawrence laboratory, using taxpayers' funds, happened to be the promoter of the Plowshare technology.

But when Tamplin was invited by community groups to tell them about the hazards of radiation as an offset to AEC sanitized education, was that regarded by Lawrence Radiation Laboratory as part of his work assignment? Oh, no; provision of the truth concerning radiation hazards was regarded as "extracurricular" and Tamplin lost his salary for each day he contributed to society. This is another facet of the AEC and Lawrence laboratory—concept of providing a "full" picture of the radiation hazards to the American taxpayers.

Tamplin's pay docked for being away week-end

And so stumble-bumble was the Lawrence Radiation Laboratory in its effort to prove to the AEC its desire to please them, they even docked Tamplin's pay for Friday, Saturday, and Sunday while he attended a national meeting of the Twelfth Science Writers' Seminar of the American Cancer Society in San Antonio, Texas. But Lawrence Radiation Laboratory doesn't work on Saturday and Sunday. When we asked the directors of the Lawrence lab how ludicrous they intended to be in their effort to suppress the truth, we must admit they did agree to reinstate Tamplin's Saturday and Sunday pay!

That particular American Cancer Society meeting is of special interest. All scientists chosen to attend were so chosen by special invitation of the Cancer Society. Any scientific laboratory in the country would, of course, be honored to have one of its staff members chosen to attend. Indeed, at any other time, we have little doubt that Tamplin would have been kissed on both cheeks by the LRL directorate and sent off to the Cancer Society meeting with a garland of roses. But to talk about radiation-induced cancer, horrors!

Repression is not complete until it is total. And the Lawrence laboratory has wasted no time or effort trying to reach that goal. After all, big brother AEC was calling for blood, and one doesn't offend the master demanding human sacrifies for the unpardonable sin of telling the true facts concerning radiation hazards. So, having taken two-thirds of Tamplin's research staff away for "budgetary reasons," they completed the task by taking three of the remaining four associates away from him. They were extremely careful to take away his associate who did reference searches and secretarial work for him. This, they obviously reasoned, would at least put a powerful crimp in any of his efforts to do further investigation of the hazard of radiation to humans. Tamplin is not a good typist, unfortunately.

And finally, Tamplin and Gofman were threatened with dismissal from Lawrence Radiation Laboratory if they persisted in what they were doing. And what crime were they committing?

- Finding out the truth about radiation hazards.
- Making the findings widely available.
- Pointing out the inaccuracies, half-truths, and outright falsehoods of AEC and its hangers-on.

Such are the anti-social, anti-human actions of science-technology and scientist-technologists imbued with their arrogant omnipotence-beliefs. They are caught up in this dynamic; they probably believe what they are doing is right—even though it destroys democracy, makes mockery of scientific freedom, and prevents saving the planet from irreversible pollution. Shall they not be held accountable as a group? As individuals?

12 The urgent need for scientific adversaries

We have pointed out how science-technology and scientists-technologists fail in their obligations to respond to the needs of society. They respond only to their own inner dynamic. The present growing environmental crisis demonstrates that science and technology have actually begun to operate to the detriment of society in several obvious ways. Atomic energy science and technology provide cogent, but by no means isolated, examples of operating to society's detriment.

It is not mysterious that this should be the case. Science and technology in modern society are well financed endeavors of government or industry, or both, acting as bedfellows. It is an axiom that the scientists follow the dollars available to support any particular branch of scientific or technological endeavor. And they learn, without evil intent, what the facts of life are with respect to success in big science and technology.

The economic health and well-being of the scientist-technologist is directly related to the continued infusion of massive numbers of dollars into his branch of science and technology. The motivation (conscious or unconscious) to discover wonderful benefits of the particular science or technology is self-evident. Indeed, it is almost inevitable that the scientist-technologist will certainly have employment, and even rise on the ladder of success, if he maintains an undying faith in the glories for society in the continual growth of his particular branch of science and technology. Criticism of such glories is hardly calculated to increase the governmental or industrial financing of the particular branch of science and technology. And, hence, criticism rarely leads to the next rung on the ladder to success.

Thus, a process of selection goes on in science-technology which assures that those who fail to perceive the glories are steadily, and unmercifully, weeded out in the selection process. And this lesson is rarely lost upon the remaining scientists-technologists. As a result, in an established field of science or technology, we must not be at all surprised at the existence of a remarkable group of think-alikes. It is no accident that criticism of the goals and effects of the science-technology area is rarely heard from within the ranks.

A particular branch of technology can be totally without meaning in fulfilling societal needs, or obviously operating against the fulfillment of critical needs of society. With a liberal infusion of dollars, scientist-technologist support of that branch of technology will be almost unanimous. For those technological programs that are meaningless, the scientists involved are no serious threat to human welfare. They at least are thereby kept employed and away from serious mischief. For those technological programs which operate against fulfillment of societal needs, the scientists and technologists do indeed represent a direct threat to society in that they are used to support the concept of omniscience and omnipotence of science and technology.

Where scientists are ineffective

By offering credibility to the proposal of an Anti-Ballistic Missile System, these men lend credibility to the concept that nuclear war is thinkable and tolerable. Where environmental deterioration threatens as a result of technological enterprise, they create the cruel illusion that science and technology will certainly be able to rescue us—a belief that is tantamount to national suicide. And for those engaged in useless or detrimental programs we suffer the additional loss of a diversion of scientific talent and manpower from meaningful, needed programs in the service of humanity.

Some scientists have spoken out against the ABM system,

war-related research, the supersonic transport, and excessive funding of NASA. Some have complained about the wrong priorities in mission-related research. And, especially recently, some scientists have sounded the alarm concerning one or another aspect of the impending or existing environmental crisis. Why have all these scientists not been more effective? One major reason is that the majority of scientists, for reasons previously described, are "hacks" who support the proposed or on-going projects of industry or government either openly or by silence. The public, therefore, assumes that the majority of the concerned and competent scientific community supports the programs, while a few "dissidents" are making noise.

However, the major reason for lack of effectiveness of those who question what industry or government is doing in technology is that it takes money and time, both in abundance, to fight city hall. The proponents are invariably well organized and well funded by government or industry or both. Quite the opposite prevails for the opponents. Moreover, legal and regulatory procedures are so structured that the opponents must necessarily present a much stronger case against a particular program than the well-funded proponents for the program. Inertia, procedure, and rules inevitably favor the technology and its promoters, even if an environmental disaster is widely evident as a result of continuation or initiation of the technological endeavor.

Shortsightedness has been a major distinction of technological endeavor, and commonly operates not only against societal needs, but even against the long-range interests of the very promoters of the endeavor. Atomic energy is an excellent illustration of a case where both societal needs and the promoter's future are adversely affected by the technology itself and its technologists.

It is evident that we are urgently in need of a mechanism for effective criticism of present day science and technology. We must learn a mechanism for articulating a new set of priorities that could begin to lead science and technology in the direction of fulfilling society's needs. It must, by now, be obvious that this will require the funding of a group of scientists (and perhaps non-scientists in collaboration) specifically for this purpose. And the history of technological enterprise teaches us that it is absolutely essential that such groups be funded in such a manner as to be completely independent of government and industry. If economic or other reprisal remains possible, no effectiveness will be achieved. We must, above all, learn to accomplish the establishment of such groups now, as a high priority.

The scientists and others who compose such critical groups must be activists in the best sense of the word. They must necessarily interact effectively with members of Congress, with activists in many fields, and with pressure groups in the country. Such association and interaction with activists, pressure groups, and the Congress can serve two important purposes. First, the scientists in such criticism groups will be aided in their own understanding and articulation of the basic needs of society. This will, in turn, aid the activist and pressure groups and the Congress in such articulation. The association will serve a further important purpose: creation of a mechanism for stimulation of public awareness of societal needs and our status with respect to their fulfillment.

One essential feature of the entire concept is that the quality of the scientific endeavor in the criticism group must be unassailable. For if the technical detail of the critical science is other than superb, the impact will be minimal. And impact is the *sine qua non* of success of the critical endeavor.

The goals of science should be questioned

An immediate task ahead is to diminish and bring into focus the unwarranted existing public and Congressional confidence in science and technology. An adversary group of scientists must demonstrate how and why science and technology are failing to meet the needs of society. Not only are they not meeting such needs, but they are seriously compounding the problems.

We believe we have demonstrated this in the case of atomic science and technology in the foregoing chapters. Indeed, since one prospect of atomic technology, peaceful or warlike, is the obliteration, slowly or rapidly, of the human species, this particular technology is especially illustrative. But only illustrative. It is important to make similar evaluations of all major technological areas, for in these times the pernicious effects of their absence of response to societal values are felt rapidly among the two hundred million people. To the extent that naive confidence of the public and Congress prevents realization that problems are being compounded—not solved—by science and technology, the hazard grows apace. Without reprisal-free criticism there is indeed but little chance that errors in technology's directions and goals will be held up for responsible examination.

Coupling science with societal needs

Destructive criticism leads only to destruction. Hence, such criticism by an adversary group of scientists can hardly contribute to the requisite re-direction of technology any more than the absence of worthwhile direction. The real purpose of serious criticism is to ask the right questions so that constructive alternative programs can be developed; programs that can provide routes to the solutions of problems of society. Blind opposition to technology is of little more merit than blind faith in its supposed infallibility.

Much of science and technology is uncoupled from our society's needs because the right questions are never asked by either one. How else is it possible to understand that in an era of technological miracles we are faced with extensive poverty and extensive unemployment? How else is it possible to understand technology's insatiable devouring of so large a fraction of societal resources to the production of military hardware that has steadily eroded the security of everyone globally?

The environmental crisis currently upon us seems unbelievable until we recognize that technology was in no serious manner concerned with its prevention and has been the major contributor to the existence of environmental problems. In the absence of scientific and technological debate and meaningful self-examination, the right questions are never asked. Indeed, as we have seen

The urgent need for scientific adversaries 233

in atomic energy, a ruthless suppression is experienced even at the first suggestion that questions of directions and goals are relevant.

Perhaps no area deserves more urgent concern and critical examination than the problem of the three faces of the Gross National Product, and these are gross national product itself, gross national power, and gross national pollution. The first two have for so long been sacred cows; the third has become a nightmare. Yet, all three are closely interrelated and clearly must eventually become self-limiting if humans are to survive their combined ravages.

It is certainly mandatory to understand why the growing GNP progressively relates less and less to an improving quality of life, and why it is likely to seriously erode such quality further than it already has. Over and over again it is found that we come into these problems in the middle of the movie. The question of electric power is a splendid example.

Somewhere in the course of our development the production of power, especially electrical power, became a major sacred cow. Obviously power utilization was associated with production of goods and services which did indeed meet some societal needs. But it does not by any means follow that ever-increasing power utilization means a better quality of life. And it is irresponsibility in the extreme to dismiss this question with, "Are you suggesting people give up their air-conditioners?" Or, "Are you suggesting we return to the caves?" These are the responses we get as a result of the absence of criticism of goals and directions. It is, at present, almost heresy to ask, "Why more power?" Or to ask, "Why must power production increase eight or ten percent per year?" The electrical utility industry has devoted itself with vigor to a studious neglect of this primary question. One dare not offend a sacred cow.

So, instead of structuring the problem of power production and considering it in relation to society's needs in depth, including the feature of survival, we start in the middle instead of at the beginning. We devote ourselves to a mad rush to determination of *how* to produce more power, rather than *why* we should. Obviously, if we have accepted the dogma that more power production is sacrosanct, we delude ourselves into asking what fuel resource will be used. And we consider secondary questions such as how much of the resource will be available with our currently projected growth in power production. If fossil fuels appear to be limited in supply, we go on to the next erroneous question—what potential fuel appears less limited?

Nuclear fission reactors contribute to gross national pollution

And this is how we brought ourselves to the current dilemma of having embarked helter-skelter upon an ill-advised, supremely hazardous program of developing nuclear fission reactors as a source of power. That it may contribute primarily to gross national product in the form of sick and dying humans with the attendant medical care requirements is only beginning to be appreciated. That it may contribute far more to gross national pollution than to a better life for anyone is becoming ever more clear. Worse yet, the agency which has committed itself to nuclear fission, the AEC, is thereby so blinded as to treat with neglect its very own alternative program, nuclear fusion, that does indeed promise unlimited power with diminished thermal and possibly absent radioactive pollution. This exemplifies the price of asking the wrong question, suppressing questions, and proceeding with mad haste in the wrong direction.

What is ludicrous is that the cost of establishment of reprisal-free scientific groups is negligible. One hundred scientists working in centers with ten scientists per group would cost fewer than $5 million per year in toto. Numerous technological areas which are in urgent need of critical scientific examination are spending, not five million dollars per year, but billions of dollars per year —to the disadvantage of, and benign neglect of, societal needs. There is no dearth of scientists and technologists to sing the praises of technology and to worship the dogmas of self-aggrandizement. These individuals will assuredly never ask critical questions concerning growth of their own technologies for the many reasons we have previously cited. They can never become the adversaries so urgently required and so painfully absent.

If we are to survive, what is needed is the establishment of groups of competent scientists who would criticize any new application of science or expansion of technology. Or more succinctly, groups of scientists who would oppose the creation of new forms of garbage while advocating means of disposing of the presently accumulated garbage. It might seem that we are suggesting an end to technological progress. Current, misguided technologists will undoubtedly leap to this point of view.

Quite the contrary, we are only suggesting that technology must not and can no longer be an end unto itself. Rather, technology must finally begin to be a significant part of the means by which society meets its ends—not its end.

Appendix

POVERTY IN THE UNITED STATES
A. R. Tamplin

In an article entitled, "Fetal and Infant Mortality and the Environment," in the December 1969 issue of the *Bulletin of the Atomic Scientists*, I presented evidence which demonstrates that the major differences in infant and fetal mortality are based upon socio-economic conditions. The purpose of this article is to show that even in the United States, a so-called affluent society, we have poverty on a grand scale.

Infant death rate and life span

The discussion of the physiological basis for the fetal and infant death rates presented in the article mentioned above is in substantial agreement with a theory of aging advanced some years ago by Hardin Jones of the University of California in Berkeley. In this theory Dr. Jones proposed that the death rate of a population at any age was dependent upon the physiological injury that the population has accumulated up to that age. He argued that one could change the rate of accumulation of new injury and thereby alter the subsequent death rate in future years. But the data indicated that the physiological injury accumulated during childhood was far more important than that accumulated during adult life. The death rate during adult life is thus primarily determined by the injury accumulated during the period of development and maturation. The earlier in life the physiological injury is accumulated, the more significant the injury. Now, quite obviously, a population that experiences a higher death rate at all ages will have a shorter average life span. For a human population, a doubling of the death rate corresponds to a reduction of eight years in the average life span. Increasing the death rate by threefold reduces the average life span by some 13 years.

Negro infant mortality double that of whites

Jones was able to demonstrate that throughout the world, there was roughly a 1:1 correspondence between infant mortality and subsequent adult death rates. This relationship is also seen in the white-non-white data of the United States. Thus the infant death rate of a population is not an isolated statistic. It defines the physiological competence of the young adults in the population and consequently, the average life span of the population. The Negro infant mortality in this country is twice that of the white population and the average life span of the Negro population is eight years less than that of the white population. The difference between the Negro death rate and the upper middle class white death rate is roughly threefold. Consequently, these whites have an average life span that is 13 years longer than the Negro population.

What is poverty?

There is no adequate definition for poverty. It is a word of the emotional side of man. A man's definition of poverty will depend upon his individual status, his background, his ambition or drive, and his degree of immorality. Ultimately, it must be conceded that defining poverty is a moral decision. On the other hand, wealth can be defined in a rather scientific manner, to wit, wealth is optimum health. Scientifically then, poverty becomes a state of relative health.

This scientific definition of wealth and poverty is given in the accompanying figure where the infant death rate in the United States is plotted as a function of a family's yearly income based upon 1965 data. This curve is drawn on the basis of 3 points: (1) The average Negro income and infant death rate, (2) the average white income and infant death rate, and (3) an assumed present-day limit to infant mortality of 12.5 per 1,000 live births plotted at $10,000. A similar curve would be obtained using the British data. The lowest income group in England has an infant death rate comparable to our Negro population. This curve demonstrates, under this scientific definition of wealth, that families with an income below $10,000 per year live in a state of relative poverty.

The dimensions of poverty

In 1965, 25% of the United States population had a family income in excess of $10,000 per year. Therefore, 75% of the population was in a state of relative poverty. In 1965, 50 per cent of the population (40 per cent of the white population and 75 per cent of the Negro population) had a family income of less than $6,000 per year. The figure indicates that incomes below this level are associated with infant death rates that are more than double that of the upper 25% group. The preceding section of this article (on life span) indicates that this infant death rate is associated with more than an eight year reduction in the average life span. The government figure for the poverty level income during this period was in the neighborhood of $3,000 per year. This income is associated with a fourfold increase in infant death rate and a 16 year reduction in life span. Some 20% of the population (17% of the white population and 37% of the Negro population) was below this governmental poverty level income. It would seem that by most definitions we have poverty in this country and have it on a grand scale. Like any other form of environmental pollution, we should seriously consider what is an acceptable level. Then, hopefully, we can exert sufficient pressure to cause our national priorities to be realigned so that our advances in science and technology, our money and knowledge, and our manpower can be put to work on improving the quality of life in our country.

The dimensions of the poverty problem as shown by the figure demonstrate that the solution of this problem should be given the highest national priority. But poverty is not simply a matter of money or jobs; it is a way of life. Poverty is a family income within a social, economic, and cultural milieu. It is the product of the numerous interplaying factors that represent the total society.

Thus, poverty is not an isolated problem. Its solution will involve most of the major problems that confront rural and urban America today. For example, the national statistics show that 50% of the girls who were in high school in the late 50's and early 60's were destined to raise families on an income below $6,000 per year. I wonder whether their education was truly relevant to this fact of life. Maybe the food industry in this country should produce, market, and advertise products such as low-cost, high-protein food supplements.

The relationship between the private automobile and smog is legend. The need for an automobile is a sizeable financial drain on low-income families. Hence, poverty is related to the problem of air pollution and the need for adequate systems of public transportation.

What we need is a master plan for improving the quality of life in this country. It cannot be a piecemeal operation. What is done in one sector affects all others. We have the capability today to look at these large problems, to isolate the various interplaying factors, to determine the nature of the interplay, and to propose an integrated solution to the problems. We have the scientific and technological knowledge; we have the industrial capability operating within an extremely viable free enterprise system; and as the space program demonstrated, we have a genius for organization. We could improve the quality of life in this country if we made the effort.

THE 0-1 YEAR DEATH RATE AS A FUNCTION OF FAMILY INCOME (1965)

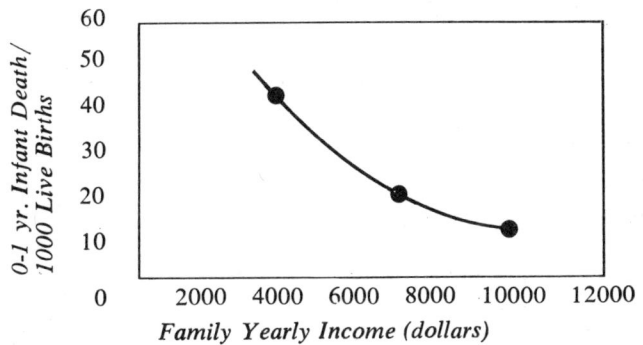

Notes

1. Joshua Lederberg, "Government is the Most Dangerous of Genetic Engineers," *Washington Post*, 19 July 1970.
2. Brian MacMahon, "The Epidemiological Aspects of Cancer," in *Cancer: A Manual for Practitioners*, ed. Thomas W. Botsford (Boston: American Cancer Society, Massachusetts Division, 1968), p. 5.
3. W. M. Court-Brown and Richard Doll, "Mortality from Cancer and Other Causes After Radiotherapy for Ankylosing Spondylitis," *Brit. Med. J.* 2, no. 5474 (4 Dec. 1965): 1327-32.
4. Alice Stewart and G. W. Kneale, "Radiation Dose Effects in Relation to Obstetric X-Rays and Childhood Cancer," *Lancet* 1, no. 7658 (6 June 1970): 1185-88.
5. United Nations, General Assembly, *Report of the Scientific Committee on the Effects of Atomic Radiation*, Suppl. 14 (A/6314), 1966, p. 127.
6. *Recommendations of the International Commission on Radiological Protection (adopted 17 Sept. 1965)*, ICRP Publ. 8 (London: Pergamon Press, 1966), p. 15.
7. United Nations, General Assembly, *Report of the Scientific Committee on the Effects of Atomic Radiation*, Suppl. 16 (A/5216), 1966, p. 85.
8. Cedric O. Carter, "Multifactorial Genetic Disease," *Hospital Practice* 5, no. 5 (May 1970): 45-9, 55-9.
9. *Report on the Effects of Atomic Radiation*, Suppl. 16 (A/5216), p. 19.
10. Mary B. Meyer et al., "Investigation of the Effects of Prenatal X-Ray Exposure of Human Oogania and Oocytes as Measured by the Later Reproductive Performance," *Am. J. of Epid.* 89, no. 6 (1969): 619-635.
11. U. S., Federal Radiation Council, *Staff Report on Guidance for the Control of Radiation Hazards in Uranium Mines*, no. 8, rev. (Washington: Government Printing Office, Sept. 1967), p. 44.
12. Umberto Saffiotti, "A Striking Synergistic Effect of the Induction of Lung Tumors" (Paper delivered at the Twelfth Science Writers' Seminar for the American Cancer Society, San Antonio, Tex., 21 Mar. 1970).
13. Werth to Gravel, 23 Dec. 1969.
14. U. S., Congress, Joint Committee on Atomic Energy, *Fallout, Radiation Standards, and Countermeasures*, 88th Cong., 1st sess., 20-27 Aug. 1963, pt. 2: 915-1075.
15. Arthur R. Tamplin and H. L. Fisher, Estimation of Dosage to Thyroids of Children in the U. S. from Nuclear Tests Conducted in Nevada During 1952 Through 1955, UCRL-14707 (Livermore, Calif.: University of California, Lawrence Radiation Laboratory, 1966).
16. *Prediction of the Maximum Dosage to Man From the Fallout of Nuclear Devices* UCRL-50163, pts. 1-6 (Livermore, Calif.: University of California, Lawrence Radiation Laboratory).
17. H. A. Tewes, "Plowshare and Radioactivity: Safety Considerations," *Nuclear News* 13, No. 5 (May 1970): 38-42.
18. U. S., Congress, Joint Committee on Atomic Energy, *Environment Effects of Producing Electric Power*, 91st Cong., 1st sess., 1969, pt. 1: 1073.
19. E. B. Lewis, "Leukemia and Ionizing Radiation," *Science* 125, no. 3255 (17 May 1957): 967-72.
20. Alice Stewart, J. Webb, and D. Hewitt, "A Survey of Childhood Malignancies," *Brit. Med. J.* 1 (1958): 1495-1508; Alice Stewart and G. W. Kneale, "Changes in the Cancer Risk Associated with Obstetric Radiography," *Lancet* 1 (1968): 104-7; idem, "Radiation Dose Effects in Relation to Obstetric X-Rays and Childhood Cancers," ibid. 1 (1970): 1185-8.
21. Brian MacMahon, "Pre-natal X-Ray Exposure and Childhood Cancer," *J. Natl. Cancer Inst.* 28 (1962): 1173-91; Brian MacMahon and H. Hutchinson, *Rev. Acta Un. Int. Cancer* 20 (1964): 1172.
22. Stewart and Kneale, "Radiation Dose Effects," pp. 1185-8.
23. U. S., Congress, Joint Committee on Atomic Energy, *Environmental Effects of Producing Electric Power*, pt. 2, 91st Cong., 2d sess., 28 Jan. 1970.
24. New York City Council open hearings, 4 Mar. 1970.
25. U. S., Congress, Senate, Committee on Public Works, Subcommittee on Air and Water Pollution, *Underground Uses of Nuclear Energy: Hearing on S. 3042*, 91st Cong., 1st sess., 18 Nov. 1969, pp. 58-76
26. An informal and unofficial discussion between the authors and Reps. Holifield and Hosmer and some staff members.
27. Herbert G. Lawson, "Nuclear Split," *Wall Street Journal*, 20 May 1970, p. 23.
28. Ernest J. Sternglass, "Infant Mortality and Nuclear Test," *Bulletin of the Atomic Scientists* 25, no. 4 (April 1969): 18-20.
29. Arthur R. Tamplin et al., "Fetal and Infant Mortality and the Environment," *Bulletin of the Atomic Scientists* 25, no. 10 (Dec. 1969): 23-9.
30. *Nucleonics Week*, 28 May 1970, p. 3.
31. John W. Gofman and Arthur R. Tamplin, "Low Dose Radiation and Can-

cer," *IEEE Transactions on Nuclear Science* NS-17, no. 1 (Feb. 1970): 1-9.
32. *A Report of Two Task Groups: I. Spatial Distribution of Radiation Dose. II. Relative Radiosensitivities of Different Tissues,* ICRP Pub. 14 (Oxford: Pergamon Press, 1969).
33. U.S., *Underground Uses of Nuclear Energy: Hearing on S. 3042,* pp. 58-76.
34. Joseph K. Wagoner et al., "Cancer Mortality Patterns Among U. S. Uranium Miners and Millers, 1950 Through 1962," *J. Natl. Cancer Inst.* 32 (1964): 787-801; Joseph K. Wagoner et al., "Radiation as the Cause of Lung Cancer Among Uranium Miners," *N. Eng. J. Med.* 273 (1965): 181-88.
35. U. S., Congress, Joint Committee on Atomic Energy, *Radiation Exposure of Uranium Miners* pts. 1 and 2, 90th Cong., 1st sess., 9 May-10 Aug. 1967.
36. Ibid., 2:1258.
37. Ibid., 1:309.
38. Ibid.
39. U. S., *Environment Effects of Producing Electric Power,* 1969, pt. 1:388.
40. Remarks prepared for *Nuclear Power and the Environment* (Proceedings of a conference sponsored by Deane C. Davis, Governor of Vermont, at the University of Vermont, Burlington, Vt., 1 Sept. 1969), p. 23. Proceedings furnished by the U. S. Atomic Energy Commission, Washington, D. C.
41. E. Eric Pochin, "The Development of the Quantitative Bases for Radiation Protection," *Brit. J. Rad.* 43, no. 507 (March 1970): 155.
42. *Nuclear Power and the Environment,* Proceedings, pp. 49-50.
43. Foreman to Tamplin, 4 Aug. 1969.
44. Merrill Eisenbud, "Standards of Radiation Protection and Their Implications to the Public Health" (Paper delivered at the Symposium on Nuclear Power and the Public, University of Minnesota, Minneapolis, Minn., 10-11 Oct. 1969).
45. U. S., *Environmental Effects of Producing Electric Power,* 1969, pt. 1:203-5.
46. Harold P. Green, "The Risk/Benefit Calculus in Nuclear Power Licensing" (Paper delivered at the Symposium on Nuclear Power and the Public, University of Minnesota, Minneapolis, Minn. 10-11 Oct. 1969).
47. Carl E. Bagge, Keynote address delivered at the 32nd Annual Meeting of the American Power Conference, Chicago, Ill., 21 Apr. 1970.
48. Walter H. Jordan, "Nuclear Energy: Benefits Versus Risks," *Physics Today* 23, no. 5 (May 1970): 38.
49. Adolph J. Ackerman, "Atomic Power —Who Looks After Public Safety," *IEEE Transactions on Aerospace and Electronic Systems* AES-5, no. 3 (May 1969): 363-75.
50. Robert L. Whitelaw, "Letter," ibid., p. 374.
51. U. S., Congress, House, *Congressional Record* 116, no. 101 (18 June 1970): p. H5841.
52. Gene Bryerton, "The Nuclear Dilemma," *Eugene* (Ore.) *Register Guard,* 7 Oct. 1969, p. 7A.
53. U. S., Atomic Energy Commission, "Report on Midland Plant Units 1 and 2," AEC News Release no. N-110, 26 June 1970.
54. *Nucleonics Week,* 14 May 1970, pp. 1-2.
55. Ibid., 4 June 1970, p. 3.
56. Bagge, keynote address.
57. Edward M. Teller, "Fast Reactors Maybe, When Will We Need Breeders? Maybe Never," *Nuclear News* 10, no. 8 (Aug. 1967): 21.
58. U. S., Atomic Energy Commission, *The Nuclear Industry, 1969* (Washington: Government Printing Office, 1969), pp. 266-7.
59. W. J. Kelleher, "Environmental Surveillance Around a Nuclear Fuel Reprocessing Installation, 1965-1967," *Radiological Health and Data Reports* 10, no. 8 (Aug. 1969): 335.
60. N. I. Sax et al., "Radioecological Surveillance of Waterways Around a Nuclear Fuels Reprocessing Plant," ibid., no. 8 (July 1969): 289-96.
61. *The Nuclear Industry, 1969,* p. 252.
62. U. S., *Underground Uses of Nuclear Energy: Hearing on S.3042,* pp. 452-3.
63. Cook to Tamplin and Gofman, 24 Feb. 1970.
64. *Recommendations of the ICRP, 17 Sept. 1965,* p. 4.
65. Donald P. Geesaman, *An Analysis of the Carcinogenic Risk From an Insoluble Alpha-Emitting Aerosol Deposited in Deep Respiratory Tissue,* UCRL-50387 (Livermore, Calif.: University of California, Lawrence Radiation Laboratory, 1968).
66. C. L. Sanders, R. C. Thompson, and W. J. Bair, "Lung Cancer: Dose Response Studies With Radionuclides," in *Inhalation Carcinogenesis* (Proceedings of the Biology Division, Oak Ridge National Laboratory Conference, Gatlinburg, Tenn., 8-11 Oct. 1969), pp. 285-303.
67. U. S., *Environmental Effects of Producing Electric Power,* 1970.
68. Donald P. Geesaman, "Plutonium and the Public Health" (Paper delivered at a symposium at the University of Colorado, Boulder, Colo., 19 Apr. 1970), pp. 10-11.

69. Tamplin to Bruner, 9 Aug. 1967.
70. Ira S. Lowry, "The Postattack Population of the United States," Rand Corp. Memorandum RM-5115-TAB (1966), pp. 122-3.
71. Freeman Dyson, "A Case for Missile Defense," *Bulletin of the Atomic Scientists* 25, no. 4 (Apr. 1969): 31-6.
72. Freeman Dyson, "Comments on Sternglass Thesis," ibid., no. 6 (June 1969): 6.
73. Arthur C. Upton, *Radiation Injury, Effects, Principles, and Perspectives* (Chicago: University of Chicago Press, 1969), p. 94.
74. Ibid.
75. Ibid., pp. 94-5.
76. Ibid., p. 95.
77. *Recommendations of the ICRP, 17 Sept. 1965*, p. 15.
78. Ibid.
79. MacMahon, "The Epidemiological Aspects of Cancer," p. 5.
80. Roger Rapoport, "U. S. Responding to Radiation Warning," *San Francisco Chronicle*, 18 Dec. 1969, p. 10.
81. Theos Thompson, "Power Technology and the Future" (Paper delivered at the Briefing Conference for State and Local Government Officials on Nuclear Development, Columbia, S. Car., 21 May 1970).
82. Lawson, "Nuclear Split," p. 23.
83. Sagan, Thomas et al. to Holifield, 30 Mar. 1970.

Internationally known for their research on the effects of radioactivity on the environment, the authors are research associates at Lawrence Radiation Laboratory in Livermore, California.

ARTHUR R. TAMPLIN is a graduate of the University of California at Berkeley, with a B.A. degree in biochemistry and a Ph.D. in biophysics. As a group leader in the Biomedical Division at Lawrence, he has been responsible for developing the ability to predict the ultimate distribution within the biosphere—particularly the concentration in man—of each radionuclide produced in the explosion of a nuclear device. This program also is concerned with the effects of its radiation on man. Previously he was a research associate with Rand Corporation, working on problems concerned with the space program.

JOHN W. GOFMAN has degrees from the University of California, a Ph.D. in nuclear-physical chemistry (at Berkeley) and an M.D. (at San Francisco). In addition to his research at Lawrence, he is a Professor of Medical Physics at Berkeley. From 1963 to 1969, he was Associate Director of the Lawrence laboratory.

Dr. Gofman is a co-discoverer of U^{232}, Pa^{232}, U^{233} and Pa^{233}; and of slow and fast neutron fissionability of U^{233}. He also is co-inventor of the uranyl acetate and columbium oxide processes for plutonium separation. He has taught in the radioisotope and radiobiology fields for over 20 years, and has done research in radiochemistry, macromolecules, lipoproteins, coronary heart disease, arteriosclerosis, trade element determination, x-ray spectroscopy, chromosomes and cancer, and radiation hazards.

Co-editor of *Advances in Biological and Medical Physics,* he also has written about 130 articles, and has prepared many additional reports. In addition, he is author, with A. V. Nichols and E. V. Dobbin, of three books on heart disease. He is a member of 10 scholastic and professional societies.

DATE DUE

~~MAR 1 1975~~			
~~DEC 17 1977~~			
~~OCT 26 1978~~			
~~DEC 28 1978~~			
~~Reserve~~			
~~Keller~~			
~~DEC 18 '79~~			
~~April 21, 1980~~			
~~OCT 26 1981~~			
~~MAY 20 1991~~			

GAYLORD — PRINTED IN U.S.A.